T0224990

SpringerBriefs in Water Science and Technology

SpringerBriefs in Water Science and Technology present concise summaries of cutting-edge research and practical applications. The series focuses on interdisciplinary research bridging between science, engineering applications and management aspects of water. Featuring compact volumes of 50 to 125 pages (approx. 20,000–70,000 words), the series covers a wide range of content from professional to academic such as:

- Literature reviews
- In-depth case studies
- Bridges between new research results
- Snapshots of hot and/or emerging topics

Topics covered are for example the movement, distribution and quality of freshwater; water resources; the quality and pollution of water and its influence on health; and the water industry including drinking water, wastewater, and desalination services and technologies.

Both solicited and unsolicited manuscripts are considered for publication in this series.

Yukun Ma · Prasanna Egodawatta · James McGree ·
Ashantha Goonetilleke

Human Health Risk Assessment of Toxic Chemical Pollutants in Stormwater

Implications for Urban Stormwater Reuse

 Springer

Yukun Ma
School of Environment
Beijing Normal University
Beijing, China

Prasanna Egodawatta
Faculty of Engineering
Queensland University of Technology
Brisbane, QLD, Australia

James McGree
Faculty of Science
Queensland University of Technology
Brisbane, QLD, Australia

Ashantha Goonetilleke
Faculty of Engineering
Queensland University of Technology
Brisbane, QLD, Australia

ISSN 2194-7244   ISSN 2194-7252 (electronic)
SpringerBriefs in Water Science and Technology
ISBN 978-981-19-9615-3   ISBN 978-981-19-9616-0 (eBook)
https://doi.org/10.1007/978-981-19-9616-0

© The Author(s), under exclusive license to Springer Nature Singapore Pte Ltd. 2023
This work is subject to copyright. All rights are solely and exclusively licensed by the Publisher, whether the whole or part of the material is concerned, specifically the rights of translation, reprinting, reuse of illustrations, recitation, broadcasting, reproduction on microfilms or in any other physical way, and transmission or information storage and retrieval, electronic adaptation, computer software, or by similar or dissimilar methodology now known or hereafter developed.
The use of general descriptive names, registered names, trademarks, service marks, etc. in this publication does not imply, even in the absence of a specific statement, that such names are exempt from the relevant protective laws and regulations and therefore free for general use.
The publisher, the authors, and the editors are safe to assume that the advice and information in this book are believed to be true and accurate at the date of publication. Neither the publisher nor the authors or the editors give a warranty, expressed or implied, with respect to the material contained herein or for any errors or omissions that may have been made. The publisher remains neutral with regard to jurisdictional claims in published maps and institutional affiliations.

This Springer imprint is published by the registered company Springer Nature Singapore Pte Ltd.
The registered company address is: 152 Beach Road, #21-01/04 Gateway East, Singapore 189721, Singapore

Preface

Urbanization results in the introduction of a range of pollutants as a result of anthropogenic activities common to urban areas. These pollutants will initially build-up on road and roof surfaces during the dry period and can wash-off with urban stormwater runoff during rainfall events. Some of these pollutants are toxic to human health such as heavy metals (HMs) and polycyclic aromatic hydrocarbons (PAHs) which have been identified as the primary toxic chemical pollutants present in urban stormwater. Urban stormwater is considered as an alternative resource to overcome the ever-increasing water scarcity being experienced in many parts of the world. However, the presence of HMs and PAHs in urban stormwater could pose risk to human health through stormwater reuse. In order to ensure the safety of stormwater reuse, it is essential to reduce the human health risk posed by HMs and PAHs present in urban stormwater. In this context, accurate assessment of the risk associated with urban stormwater is essential for the formulation of effective stormwater management strategies.

This monograph focuses on human health risk assessment of HMs and PAHs present in urban stormwater. Since traffic and land use are the primary sources of HMs and PAHs in the urban environment, this research study selected 20 study sites with a range of traffic and land use characteristics in Gold Coast, Queensland State, Australia. As stormwater risk is dependent on pollutant build-up, build-up samples on road surfaces at the study sites were collected through field investigations. The concentration of HMs and PAHs in the build-up was determined through laboratory analysis. Multivariate data analysis techniques and mathematical modeling approaches were employed to create new knowledge on the influence of traffic and land use characteristics on stormwater risk and to investigate the mathematical relationships between risk and traffic and land-use-related factors.

It was found that HM and PAH concentrations in stormwater and the associated risk were the highest in industrial areas, followed by commercial and residential areas, and lowest in natural areas. For the same land use, human health risk posed by HMs and PAHs in urban stormwater generally increases with increasing traffic volume. Additionally, motor vehicle-related activities and brake–start activities also exert significant influences on stormwater risk. Accordingly, a series of mathematical equations were developed to define the human health risk posed by HMs and PAHs in urban stormwater based on traffic and land use characteristics, including daily traffic volume (DTV), braking and starting frequency (BSF), commercial, industrial, and residential area percentages (C, I, and R), and the fraction of vehicle-related businesses (FVS). This approach constitutes a novel stormwater risk assessment approach.

Based on the risk assessment procedure derived in this research study, Cr and heavy PAHs (PAHs with 5–6 benzene rings) were identified as the most toxic HM and PAH species present in urban stormwater and contributing the largest fraction of risk to human health. This suggests that stormwater management should focus on the elimination of Cr and heavy PAHs in the urban environment and the use of alternate materials or technologies that do not contain Cr and heavy PAHs. In addition, the mathematical equations developed to define the relationships between risk and influential factors can also be employed as an urban planning tool for stormwater management in the context of changing traffic and land use characteristics in order to identify priority areas with high risk. Appropriate engineering control techniques such as water sensitive urban design (WSUD) can be applied in the priority areas to reduce stormwater risk. Administrative control and personal protective equipment are the other approaches that can also be applied in the priority areas to minimize the risk to human health from polluted stormwater.

The outcomes derived from this study confirm that only evaluating the risk from individual HM and PAH concentrations in stormwater cannot provide comprehensive confirmation of safety in relation to stormwater reuse. The study found that even though an individual HM and PAH may not be toxic to human health, multiple HMs and PAHs mixtures can pose risk to human health. This indicates that taking into consideration the combined toxic effects of multiple HMs and PAHs is essential for effective stormwater quality management. This is contrary to the guidance provided in current water quality guidelines.

Beijing, China   Yukun Ma
Brisbane, Australia   Prasanna Egodawatta
Brisbane, Australia   James McGree
Brisbane, Australia   Ashantha Goonetilleke

Contents

Chapter 1
Human Health Risk from Stormwater Pollution

Abstract Urban stormwater contains a diversity of pollutants and some can be toxic to human health such as heavy metals (HMs) and polycyclic aromatic hydrocarbons (PAHs). HMs and PAHs are primarily generated from traffic and land use related activities. They can pose potential risks to human health through stormwater reuse for recreational and potable purposes. In order to ensure safety in stormwater reuse, risk minimization is essential. Accurate and efficient assessment of the risk posed by HMs and PAHs in urban stormwater can provide a robust foundation for stormwater risk management in the context of stormwater reuse.

Keywords Stormwater quality · Heavy metals · Polycyclic aromatic hydrocarbons · Traffic characteristics · Land use activities · Stormwater reuse · Risk assessment

1.1 Background

Urbanization results in an increase in anthropogenic activities such as traffic and land use related activities which can introduce large loads of pollutants to the urban environment. Heavy metals (HMs) and polycyclic aromatic hydrocarbons (PAHs) are the primary toxic pollutants generated from traffic and land use related activities (Ma et al., 2017a, 2017b, 2017c). After being emitted from a range of sources, these pollutants will initially deposit on urban surfaces such as roads and roofs, and can wash-off with stormwater runoff during rainfall events. Urban stormwater reuse is becoming increasingly prevalent to overcome the serious water scarcity around the world (Managi et al., 2016; Rodriguez-Narvaez et al., 2021; Wijesiri et al., 2020). However, the presence of HMs and PAHs in stormwater could pose risk to human health through reusing stormwater (Ma et al., 2017a, 2017b, 2017c). For example, HMs can exert acute and chronic toxic effects on humans such as hypertension, renal dysfunction and Parkinson's disease, while the main concern in relation to the toxic effects of PAHs is their potential to pose cancer risk (Nubi et al., 2022).

In order to guarantee the safety of urban stormwater reuse, it is necessary to mitigate the associated potential human health risks. Stormwater risk management requires accurate assessment of the risk posed by HMs and PAHs in urban stormwater.

© The Author(s), under exclusive license to Springer Nature Singapore Pte Ltd. 2023

Y. Ma et al., *Human Health Risk Assessment of Toxic Chemical Pollutants in Stormwater*, SpringerBriefs in Water Science and Technology, https://doi.org/10.1007/978-981-19-9616-0_1

Stormwater risk is dependent on stormwater quality which is influenced by a range of source factors such as traffic volume, traffic conditions, land use related activities and land use characteristics (Goonetilleke & Lampard, 2019; Wijesiri et al., 2018). In this context, an efficient and accurate estimation of the human health risk posed by HMs and PAHs in urban stormwater is critical which entails the appropriate consideration of the relevant influential factors. Therefore, a fit-for-purpose stormwater risk management strategy is essential for enhancing urban stormwater reuse.

1.2   Urban Stormwater Pollution

1.2.1   Primary Stormwater Toxic Chemical Pollutants

1.2.1.1   Heavy Metals

HMs are among the most common chemical pollutants in urban stormwater. The most common HMs found in urban stormwater include aluminum (Al), cadmium (Cd), chromium (Cr), copper (Cu), iron (Fe), lead (Pb), manganese (Mn), nickel (Ni) and zinc (Zn) (Gunawardana et al., 2015). Cr and Pb are extremely toxic to people even if present at trace level. They can lead to serious illness such as hypertension and renal dysfunction (Al-Saleh et al., 2017). Other metals are relatively less toxic compared to Cr and Pb, but can also cause adverse impacts on humans after long exposure. For example, accumulation of Al in the body can adversely impact the nervous system while Mn can lead to Parkinson's disease (Raj et al., 2021). Cr and Ni can exert adverse effects on the skin and can result in allergies (Vandana et al., 2022). Cu, Fe and Zn are essential elements for the human body, but excessive exposure to these metals can cause nausea, vomiting and damage to organs (Zoroddu et al., 2019).

Intensive traffic and land use related activities can lead to high concentrations of HMs in stormwater. For example, high traffic volume and frequent traffic congestion common in industrial and commercial areas can generally lead to high concentrations of HMs in stormwater (Liu et al., 2016). Besides, the dominant vehicle types are different in different land use areas (Gunawardena et al., 2012). Trucks are more prevalent in industrial areas while cars are more predominant in commercial and residential areas. This leads to the different types and loads of HMs from different land uses. Diesel fuel used in trucks mainly generate high loadings of Cr, Cu, Ni, Pb and Zn, while gasoline used in cars can result in high emission of Cd.

1.2.1.2   Polycyclic Aromatic Hydrocarbons

PAHs are among the main toxic chemical pollutants found in urban stormwater due to their prevalence in the urban environment and exert carcinogenic effects on human health (Esfandiar et al., 2021; Gbeddy et al., 2021). The presence of these compounds can lead to cancer risk to the inhalation system, nervous system and renal system in the human body (Verma et al., 2022). PAHs can be divided into light PAHs (PAHs containing two to four benzene rings) and heavy PAHs (PAHs containing greater than four benzene rings) (Pathak et al., 2022). USEPA has identified 16 PAHs as priority toxic pollutants. Among these, light PAHs include naphthalene (NAP), acenaphthylene (ACY), acenaphthene (ACE), fluorene (FLU), phenanthrene (PHE), anthracene (ANT), fluoranthene (FLA), pyrene (PYR), benz[a]anthracene (B[a]A), chrysene (CHR), while heavy PAHs include benzo[b]fluoranthene (B[b]F), Benzo[k]fluoranthene (B[k]F), benzo[a]pyrene (B[a]P), indeno(1,2,3-cd)pyrene (IND), dibenzo[a,h]anthracene (D[a]A), and benzo[ghi]perylene (B[g]P) (USEPA, 1984).

Exhaust emissions is a major source of PAHs on urban road surfaces (Mummullage et al., 2016a, 2016b). Therefore, a high traffic volume generally leads to high loadings of PAHs in stormwater (Wei et al., 2015). Additionally, different vehicles types contribute different PAHs species. Diesel combustion primarily generates light PAHs while gasoline combustion mainly generates heavy PAHs (Gunawardena et al., 2012). As such, stormwater from industrial areas contains more light PAHs and heavy PAHs are more dominant in stormwater from commercial and residential areas. Land use also influences PAH concentrations in stormwater (Gbeddy et al., 2022; Goonetilleke et al., 2005). For example, stormwater from industrial areas usually contains higher concentrations of PAHs due to the nature of the industrial activities and the intense traffic generated by the associated anthropogenic activities.

1.2.2 Toxic Chemical Pollutant Sources

1.2.2.1 Traffic Related Activities

Traffic related activities are a major source of HMs and PAHs in the urban environment (Li et al., 2018). Roadside soil and road surface abrasion, tyre wear and brake wear have been identified as the key traffic related sources of HMs in urban road dust (Gunawardena et al., 2015; Mummullage et al., 2016a, 2016b). Roadside soil and road surface abrasion by moving vehicles on roads is the largest contributor of HMs in build-up. Al, Fe, Cr, Mn and Pb are mainly generated from roadside soil and road surface abrasion. Tyre wear contributes large amounts of Zn to build-up, while brake wear is the major source of Cd and Cu. PAHs are primarily contributed by vehicle exhaust due to incomplete fuel combustion (Mummullage et al., 2016a, 2016b). PAHs tend to be associated with particles, and the particles generated from exhaust emissions are normally ultrafine (<0.1 μm) (Moreno-Ríos et al., 2022). Therefore, exhaust emitted PAHs will initially remain in the atmosphere and deposit on urban surfaces via dry or wet deposition (Wei et al., 2020).

1.2.2.2   Land Use Related Activities

Urban land use related activities are also an important source of HMs and PAHs to road dust and stormwater runoff (Li et al., 2022). Industrial activities, such as metal industry, manufacturing operations and heavy-duty vehicle movements, can generate large amounts of particulates, HMs and PAHs. Besides, road surfaces in industrial areas is usually in poor condition which can contribute high loadings of pollutants. Commercial activities such as fuel stations and vehicle repair facilities can introduce large amounts of HMs and PAHs due to the nature of the activities associated with these facilities. Additionally, intensive commercial activities, such as retail, entertainment, office premises and hospitality can attract high traffic volumes and various traffic related activities (Ziyath et al., 2016). Consequently, the high traffic volume and congested traffic can become another critical source of HMs and PAHs in commercial areas. Activities associated with residential areas, especially gardening, is a major source of organic matter. Organic matter adhering to particulate surfaces is able to enhance the adsorption of HMs and PAHs to solids (Jayarathne et al., 2019a, 2019b; Miranda et al., 2021). During rainfall events, HMs and PAHs associated with solids deposited on urban road and roof surfaces are transported by stormwater runoff (Gunawardana et al., 2014). Consequently, high organic matter coating on solids can exacerbate the pollution of HMs and PAHs in urban stormwater.

1.3   Risk Assessment in Relation to Toxic Chemical Pollutants

1.3.1   General Concepts of Risk

The definition of risk generally comprises of two components, namely, the detrimental effects of hazards and its likelihood of occurrence (USEPA, 1989). The detrimental effects refer to damage or injury resulting from hazards and the likelihood of occurrence is the frequency of the detrimental consequences. Hazards are identified as physical, chemical, biological or radiological substances that can exert harmful impacts to humans or the environment. Some hazards can cause acute harm over a short period while some hazards can lead to chronic injuries over a long period of time. The risk posed by hazards needs mitigation to ensure a safe environment for humans and the environment. Assessment of risk is essential for risk management and it is defined as an approach to provide knowledge for risk management decision making.

1.3.2   Risk Assessment Procedure

Risk assessment consists of four steps, namely, hazard identification, exposure assessment, dose-response assessment and risk characterization, as illustrated in Fig. 1.1 (USEPA, 1989). Hazard identification is the evaluation of the toxic effects of the hazards. Exposure assessment is the identification of the locations, pathways, concentrations and intake of the hazards. Dose-response assessment is the evaluation of the likelihood of the detrimental effects of the hazards. Risk characterization is the quantification of the risk based on exposure assessment and dose-response assessment.

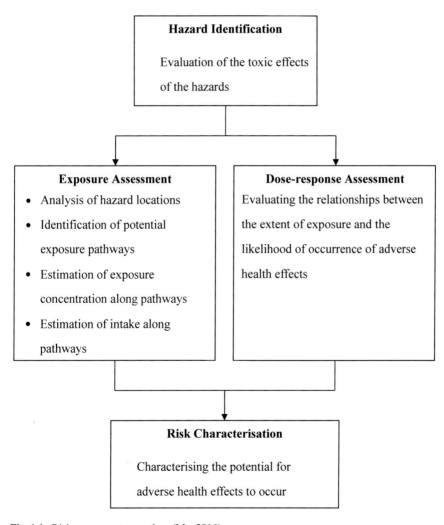

Fig. 1.1   Risk assessment procedure (Ma, 2016)

1.3.2.1 Hazard Identification

Hazard identification is assessing whether a substance can lead to adverse impacts on human or ecosystem health. The potential toxic effects of the substance also need identification in this step. HMs and PAHs have been identified as common toxic chemical pollutants commonly present in urban stormwater. The discussion below is on human health as this is the focus of this work. Generally, toxic substances can either have non-carcinogenic or carcinogenic effects on human health.

1.3.2.2 Exposure Assessment

There are three steps involved in exposure assessment: (1) to characterize the locations of the hazards; (2) to identify the exposure pathways through which people can come into contact with the hazards and the relevant exposure mechanisms; (3) to determine the intake of hazards by people through each pathway and specific frequency of exposure and duration. The intake is generally quantified according to Eq. 1.1.

$$Intake = \frac{C \times CR \times EF \times ED}{BW} \times \frac{1}{AT} \qquad (1.1)$$

where

$Intake$ intake of the hazards by people (mg/(kg day));
C concentration of a hazard (e.g. mg/L);
CR contact rate by people (e.g. L/day);
EF exposure frequency (days/year);
ED exposure duration (years);
BW average body weight of the exposed people (kg);
AT average time period of exposure (days).

1.3.2.3 Dose-response Assessment

Dose-response is the step to assess the relationship between the exposure dose and the probability of detrimental response to hazards. As discussed under hazard identification, HMs and PAHs can pose either non-carcinogenic or carcinogenic risk to human health. In assessing the non-carcinogenic effects of the hazards, the dose-response relationship can be characterized as a dose-response curve. Figure 1.2 provides an example of a dose-response curve which shows that an exposure dose under 30 mg/kg is safe for human health, while a dose higher than 30 mg/kg would result in a response to the hazard (Note: This is an example only to demonstrate the use of a dose-response curve). The threshold dose is normally considered as either the largest dose under which the hazard is safe to human health (i.e. the No Observed Adverse Effects Level or NOAEL) or the smallest dose above which the hazards can pose risk to human

Fig. 1.2 An example of a dose-response curve

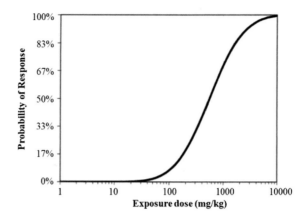

health (i.e. the Lowest Observed Adverse Effect Level or LOAEL). This threshold dose is referred to as reference dose (RfD) for assessing non-carcinogenic risk to human health (USEPA, 1989). The reference dose represents the maximum dose of the hazard that can be exposed to people. Therefore, the reference dose of non-carcinogenic pollutants in urban stormwater should be identified in dose-response assessment.

In terms of estimating the carcinogenic effects of a hazard, the cancer slope factor (CSF) is usually adopted to characterize the cancer risk (USEPA, 2005). CSF is the gradient of the dose-response curve and it is defined as the upper-bound of the likelihood to cause cancer. The dose-response curve for cancer risk is generally hypothesized to be linear such that the CSF is a constant value (see Fig. 1.3). Therefore, the CSF for the carcinogenic pollutants in urban stormwater should also be identified in the dose-response assessment.

Fig. 1.3 Cancer slope factor for carcinogenic hazards (Ma, 2016)

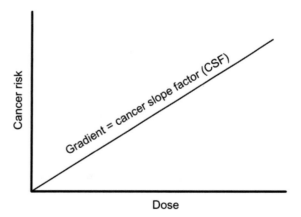

1.3.2.4   Risk Characterization

Risk characterization gives either qualitative or quantitative results for the risk assessment. The non-carcinogenic risk is estimated by comparing the exposure dose of a hazard with the RfD of the hazard, and the resulting quotient is characterized as hazard quotient (HQ) (see Eq. 1.2) (USEPA, 1989). If the intake is less than the corresponding RfD, the hazard would not exert toxic effects on human health. If the intake is greater than the RfD, people would be susceptible to the hazard.

$$HQ = \frac{E}{RfD} \tag{1.2}$$

where

E exposure level (mg/(kg day));
RfD reference dose (mg/(kg day)).

The carcinogenic risk is estimated using an incremental lifetime cancer risk (ILCR) (USEPA, 2005). It is defined as the incremental probability of people having cancer during their lifetime. Cancer risk is normally determined by multiplying the exposure level through a specific pathway with the corresponding CSF according to Eq. 1.3. If ILCR is less than the threshold probability, the carcinogenic effect on human health is considered negligible. Otherwise, the carcinogenic effect on human health is considered unacceptable.

$$Risk = E \times CSF \tag{1.3}$$

where

E exposure level (mg/(kg day));
CSF cancer slope factor ((mg/(kg day))$^{-1}$).

1.4   Rationale for the Monograph

The main objective of this monograph is to contribute new knowledge on human health risk posed by HMs and PAHs in urban stormwater and for developing mathematical relationships between urban stormwater risk and traffic and land use related activities, and accordingly, to provide the way forward for urban stormwater risk management and stormwater reuse for recreational and potable purposes. This objective was achieved based on a robust methodology consisting of three phases: (1) field investigations to collect data in relation to common traffic and land use generated toxic chemical pollutants, namely, HMs and PAHs; (2) data analysis in relation to urban stormwater quality estimation for the identification of the impacts of traffic and land use factors on HM and PAH concentrations in stormwater; (3) development of mathematical relationships for stormwater risk assessment based on traffic and land use variables.

The outline of the research study undertaken is illustrated in Fig. 1.4. Chapter 1 provides a state-of-the-art literature review in relation to the research study and this provided the foundation for the other chapters. Chapter 2 presents the research design for the comprehensive field investigations and laboratory analyses undertaken. Chapter 3 discusses the application of the field and laboratory data derived from Chap. 2 to estimate HM and PAH concentrations in stormwater and the identification of the relationships between HM and PAH concentrations with traffic and land use factors. Chapter 4 presents the development of the risk assessment models based on pollutant concentration data derived from Chap. 3 to predict human health risks posed by HMs and PAHs in stormwater. Chapter 5 discusses practical application of the study outcomes for stormwater risk management and recommendations for future research directions.

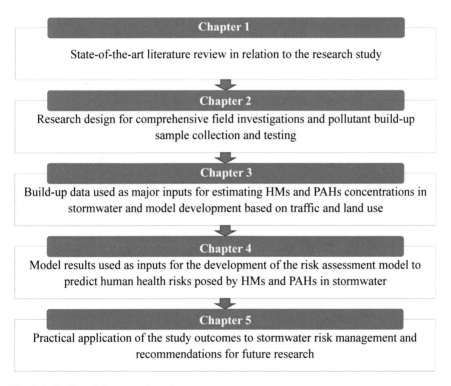

Chapter 1

State-of-the-art literature review in relation to the research study

Chapter 2

Research design for comprehensive field investigations and pollutant build-up sample collection and testing

Chapter 3

Build-up data used as major inputs for estimating HMs and PAHs concentrations in stormwater and model development based on traffic and land use

Chapter 4

Model results used as inputs for the development of the risk assessment model to predict human health risks posed by HMs and PAHs in stormwater

Chapter 5

Practical application of the study outcomes to stormwater risk management and recommendations for future research

Fig. 1.4 Outline of the research study

1.5 Summary

This chapter has provided a focused literature review on major traffic and land use generated toxic chemical pollutants in urban stormwater and the assessment of risk posed by these toxic pollutants. HMs and PAHs have been identified as the main toxic chemical pollutants in urban stormwater. These compounds can pose risk to human health if people come into contact with stormwater. The main sources of HMs and PAHs in urban stormwater include traffic related activities, such as roadside soil erosion and road surface abrasion, brake wear, tyre wear and by-products of fossil fuels combustion, and a range of urban land use related activities, such as commercial, industrial and residential activities.

Urban stormwater can be reused as a resource to overcome the challenge posed by the escalating scarcity of freshwater resources worldwide. This emphasizes the necessity to reduce human health risk posed by the toxic pollutants in stormwater. Accurate and efficient assessment of the risk is critical for stormwater risk management.

References

Al-Saleh, I., Al-, R., Elkhatib, R., Abduljabbar, M., & Al-, T. (2017). Risk assessment of environmental exposure to heavy metals in mothers and their respective infants. *International Journal of Hygiene and Environmental Health, 220*(8), 1252–1278.

Esfandiar, N., Suri, R., & McKenzie, E. R. (2021). Simultaneous removal of multiple polycyclic aromatic hydrocarbons (PAHs) from urban stormwater using low-cost agricultural/industrial byproducts as sorbents. *Chemosphere, 274*, 129812.

Gbeddy, G., Egodawatta, P., Akortia, E., & Goonetilleke, A. (2022). Inherent and external factors influencing the distribution of PAHs, hydroxy-PAHs, carbonyl-PAHs and nitro-PAHs in urban road dust. *Environmental Pollution, 308*, 119705.

Gbeddy, G., Egodawatta, P., Goonetilleke, A., Akortia, E., & Glover, E. T. (2021). Influence of photolysis on source characterization and health risk of polycyclic aromatic hydrocarbons (PAHs), and carbonyl-, nitro-, hydroxy- PAHs in urban road dust. *Environmental Pollution, 269*, 116103.

Goonetilleke, A., & Lampard, J.-L. (2019). Chapter 3 - Stormwater quality, pollutant sources, processes, and treatment options. In A. K. Sharma, T. Gardner, & D. Begbie (Eds.), *Approaches to water sensitive urban design* (pp. 49–74). Woodhead Publishing.

Goonetilleke, A., Thomas, E., Ginn, S., & Gilbert, D. (2005). Understanding the role of land use in urban stormwater quality management. *Journal of Environmental Management, 74*(1), 31–42.

Gunawardana, C., Egodawatta, P., & Goonetilleke, A. (2014). Role of particle size and composition in metal adsorption by solids deposited on urban road surfaces. *Environmental Pollution, 184*, 44–53.

Gunawardana, C., Egodawatta, P., & Goonetilleke, A. (2015). Adsorption and mobility of metals in build-up on road surfaces. *Chemosphere, 119*, 1391–1398.

Gunawardena, J., Egodawatta, P., Ayoko, G. A., & Goonetilleke, A. (2012). Role of traffic in atmospheric accumulation of heavy metals and polycyclic aromatic hydrocarbons. *Atmospheric Environment, 54*, 502–510.

Gunawardena, J., Ziyath, A. M., Egodawatta, P., Ayoko, G. A., & Goonetilleke, A. (2015). Sources and transport pathways of common heavy metals to urban road surfaces. *Ecological Engineering, 77*, 98–102.

Jayarathne, A., Egodawatta, P., Ayoko, G. A., & Goonetilleke, A. (2018). Assessment of ecological and human health risks of metals in urban road dust based on geochemical fractionation and potential bioavailability. *Science of the Total Environment, 635*, 1609–1619.

Jayarathne, A., Egodawatta, P., Ayoko, G. A., & Goonetilleke, A. (2019a). Transformation processes of metals associated with urban road dust: A critical review. *Critical Reviews in Environmental Science and Technology, 49*(18), 1675–1699.

Jayarathne, A., Mummullage, S., Gunawardana, C., Egodawatta, P., Ayoko, G. A., & Goonetilleke, A. (2019b). Influence of physicochemical properties of road dust on the build-up of hydrocarbons. *Science of the Total Environment, 694*, 133812.

Li, F.-J., Yang, H.-W., Ayyamperumal, R., & Liu, Y. (2022). Pollution, sources, and human health risk assessment of heavy metals in urban areas around industrialization and urbanization-Northwest China. *Chemosphere, 308*, 136396.

Li, Y., Jia, Z., Wijesiri, B., Song, N., & Goonetilleke, A. (2018). Influence of traffic on build-up of polycyclic aromatic hydrocarbons on urban road surfaces: A Bayesian network modelling approach. *Environmental Pollution, 237*, 767–774.

Liu, A., Egodawatta, P., Guan, Y., & Goonetilleke, A. (2013). Influence of rainfall and catchment characteristics on urban stormwater quality. *Science of the Total Environment, 444*, 255–262.

Liu, A., Gunawardana, C., Gunawardena, J., Egodawatta, P., Ayoko, G. A., & Goonetilleke, A. (2016). Taxonomy of factors which influence heavy metal build-up on urban road surfaces. *Journal of Hazardous Materials, 310*, 20–29.

Ma, B., Han, Y., Cui, S., Geng, Z., Li, H., & Chu, C. (2020). Risk early warning and control of food safety based on an improved analytic hierarchy process integrating quality control analysis method. *Food Control, 108*, 106824.

Ma, Y. (2016). *Human health risk of toxic chemical pollutants generated from traffic and land use activities*. Queensland University of Technology.

Ma, Y., Egodawatta, P., McGree, J., Liu, A., & Goonetilleke, A. (2016b). Human health risk assessment of heavy metals in urban stormwater. *Science of the Total Environment, 557–558*, 764–772.

Ma, Y., Liu, A., Egodawatta, P., McGree, J., & Goonetilleke, A. (2017a). Assessment and management of human health risk from toxic metals and polycyclic aromatic hydrocarbons in urban stormwater arising from anthropogenic activities and traffic congestion. *Science of the Total Environment, 579*, 202–211.

Ma, Y., Liu, A., Egodawatta, P., McGree, J., & Goonetilleke, A. (2017b). Quantitative assessment of human health risk posed by polycyclic aromatic hydrocarbons in urban road dust. *Science of the Total Environment, 575*, 895–904.

Ma, Y., McGree, J., Liu, A., Deilami, K., Egodawatta, P., & Goonetilleke, A. (2017c). Catchment scale assessment of risk posed by traffic generated heavy metals and polycyclic aromatic hydrocarbons. *Ecotoxicology and Environmental Safety, 144*, 593–600.

Managi, S., Goonetilleke, A., & Wilson, C. (2016). Water resources: Embed stormwater use in city planning. *Nature, 532*(7597), 37.

Miranda, L. S., Deilami, K., Ayoko, G. A., Egodawatta, P., & Goonetilleke, A. (2022). Influence of land use class and configuration on water-sediment partitioning of heavy metals. *Science of the Total Environment, 804*, 150116.

Miranda, L. S., Wijesiri, B., Ayoko, G. A., Egodawatta, P., & Goonetilleke, A. (2021). Water-sediment interactions and mobility of heavy metals in aquatic environments. *Water Research, 202*, 117386.

Moreno-Ríos, A. L., Tejeda-Benítez, L. P., & Bustillo, C. F. (2022). Sources, characteristics, toxicity, and control of ultrafine particles: An overview. *Geoscience Frontiers, 13*(1), 101147.

Mummullage, S., Egodawatta, P., Ayoko, G. A., & Goonetilleke, A. (2016a). Sources of hydrocarbons in urban road dust: Identification, quantification and prediction. *Environmental Pollution, 216*, 80–85.

Mummullage, S., Egodawatta, P., Ayoko, G. A., & Goonetilleke, A. (2016b). Use of physicochemical signatures to assess the sources of metals in urban road dust. *Science of the Total Environment, 541*(3), 1303.

Nubi, A. O., Popoola, S. O., Dada, O. A., Oyatola, O. O., Unyimadu, J. P., Adekunbi, O. F., & Salami, A. M. (2022). Spatial distributions and risk assessment of heavy metals and PAH in the southwestern Nigeria coastal water and estuaries, Gulf of Guinea. *Journal of African Earth Sciences, 188*, 104472.

Pathak, S., Sakhiya, A. K., Anand, A., Pant, K. K., & Kaushal, P. (2022). A state-of-the-art review of various adsorption media employed for the removal of toxic Polycyclic aromatic hydrocarbons (PAHs): An approach towards a cleaner environment. *Journal of Water Process Engineering, 47*, 102674.

Raj, K., Kaur, P., Gupta, G. D., & Singh, S. (2021). Metals associated neurodegeneration in Parkinson's disease: Insight to physiological, pathological mechanisms and management. *Neuroscience Letters, 753*, 135873.

Rodriguez-Narvaez, O. M., Goonetilleke, A., & Bandala, E. (2021). Chapter 23—Treatment technologies for stormwater reuse. In J. Rodrigo-Comino (Ed.), *Precipitation* (pp. 521–549). Elsevier.

USEPA. (1984). Guidelines establishing test procedures for the analysis of pollutants under Clean Water Act: Method 610—polynuclear aromatic hydrocarbons. Washington, *DC, US Environmental Protection Agency, 49*, 43344–43352.

USEPA. (1989). Risk assessment guidance for superfund, Volume I. Human health evaluation manual (Part A). *Washington, DC, US Environmental Protection Agency.*

USEPA. (2005). Guidelines for Carcinogen Risk Assessment. *Washington, DC, US Environmental Protection Agency.*

Vandana, P., Mahto, U., & Das, S. (2022). Chapter 2—Mechanism of toxicity and adverse health effects of environmental pollutants. In S. Das & H. R. Dash (Eds.), *Microbial Biodegradation and Bioremediation* (2nd ed., pp. 33–53). Elsevier.

Verma, P. K., Sah, D., Satish, R., Rastogi, N., Kumari, K. M., & Lakhani, A. (2022). Atmospheric chemistry and cancer risk assessment of polycyclic aromatic hydrocarbons (PAHs) and Nitro-PAHs over a semi-arid site in the Indo-Gangetic plain. *Journal of Environmental Management, 317*, 115456.

Wei, C., Bandowe, B. A. M., Han, Y., Cao, J., Zhan, C., & Wilcke, W. (2015). Polycyclic aromatic hydrocarbons (PAHs) and their derivatives (alkyl-PAHs, oxygenated-PAHs, nitrated-PAHs and azaarenes) in urban road dusts from Xi'an, Central China. *Chemosphere, 134*, 512–520.

Wei, T., Wijesiri, B., Li, Y., & Goonetilleke, A. (2020). Particulate matter exchange between atmosphere and roads surfaces in urban areas. *Journal of Environmental Sciences, 98*, 118–123.

Wijesiri, B., Deilami, K., & Goonetilleke, A. (2018). Evaluating the relationship between temporal changes in land use and resulting water quality. *Environmental Pollution, 234*, 480–486.

Wijesiri, B., Liu, A., & Goonetilleke, A. (2020). Impact of global warming on urban stormwater quality: From the perspective of an alternative water resource. *Journal of Cleaner Production, 262*, 121330.

Ziyath, A. M., Egodawatta, P., & Goonetilleke, A. (2016). Build-up of toxic metals on the impervious surfaces of a commercial seaport. *Ecotoxicology and Environmental Safety, 127*, 193–198.

Zoroddu, M. A., Aaseth, J., Crisponi, G., Medici, S., Peana, M., & Nurchi, V. M. (2019). The essential metals for humans: A brief overview. *Journal of Inorganic Biochemistry, 195*, 120–129.

Chapter 2
Research Program

Abstract Stormwater risk is primarily determined by the pollutant concentrations present, which is in turn dependent on the build-up of pollutants on urban surfaces. Therefore, risk assessment in relation to stormwater requires a robust program of research investigations into pollutant concentrations in build-up on urban surfaces. This chapter discusses the critical details of the research program as formulated, specifically in relation to study sites selection, pollutant build-up sample collection and laboratory testing.

Keywords Study site selection · Build-up sample collection · Build-up sample testing · Stormwater quality · Stormwater pollutant processes

2.1 Background

Toxic chemical pollutants such as heavy metals (HMs) and polycyclic aromatic hydrocarbons (PAHs) in urban stormwater can pose risk to human health through stormwater reuse for recreational and potable purposes (Ma et al., 2017a, 2017b). Therefore, human health risk assessment due to the presence of HMs and PAHs in urban stormwater is essential to ensure its safe reuse. An appropriate site selection protocol, reliable sample collection and an accurate sample testing program are important steps in providing a robust database to assess the potential human health risks in relation to stormwater reuse. Risks posed by HMs and PAHs are determined by their concentrations in stormwater. In turn, the concentrations of HMs and PAHs in stormwater are influenced by their build-up load (Jayarathne et al., 2018). Further, the build-up of HMs and PAHs is significantly impacted by the surrounding traffic and land use characteristics (Gunawardena et al., 2014). Consequently, data collection needed to consider a range of traffic and land use characteristics to represent the spectrum of HM and PAH concentrations present in stormwater originating from a typical urban area.

Accordingly, this chapter outlines the design of the comprehensive program of research undertaken to investigate the concentrations of HMs and PAHs in build-up samples on urban impervious surfaces. Road and roof surfaces are the most common

© The Author(s), under exclusive license to Springer Nature Singapore Pte Ltd. 2023
Y. Ma et al., *Human Health Risk Assessment of Toxic Chemical Pollutants in Stormwater*,
SpringerBriefs in Water Science and Technology,
https://doi.org/10.1007/978-981-19-9616-0_2

impervious surfaces present in an urban area. Stormwater runoff from these surfaces is primarily responsible for the transport of pollutants associated with urban areas. Critical details of the program of research relating to study sites identification, field sampling program and laboratory testing techniques are presented in this chapter. The outcomes of this work provided the necessary database for the subsequent analysis and interpretation.

2.2 Study Sites

Human health risks in urban stormwater are primarily influenced by traffic and land use activities (Ma et al., 2017a, 2017b). Therefore, typical urban road sites with varying traffic and land use characteristics were selected. Chapter 3 discusses how pollutant loads from roof surfaces were assessed based on the field data collected. The study sites were located at Gold Coast, Queensland State, Australia. A total of 20 study sites were selected spread over five urban suburbs, namely, Suburb 1, Suburb 2, Suburb 3, Suburb 4 and Suburb 5. These suburbs represented different land use types, namely, commercial, industrial, residential and natural. Four road sites were selected from each suburb to represent different traffic volumes. Figure 2.1 shows the distribution of the selected study sites. Table 2.1 gives their traffic and land use characteristics.

2.3 Sample Collection

The pollutant build-up samples were collected at the selected road study sites. Two samples were collected from each site in order to take into consideration the possible difference in pollutant loads due to variations in the antecedent dry period leading up to the sampling and the average data was used for the subsequent data analyses. Sample collection was conducted after seven to nine antecedent dry days after the preceding rainfall event. This was to allow for the fact that the pollutant load on a road surface approaches a near constant value around this antecedent dry period (Egodawatta et al., 2007). The sampling plot at each selected road site was a one meter wide rectangular area and the length was half width of the road (Fig. 2.2). This approach was adopted in view of the variation in particle size distribution of solids across the road surface (Gunawardana et al., 2014). It ensured that the build-up sample collected was representative of the pollutant load present on the road surface being sampled.

Samples were collected using the dry and wet vacuuming method using a Delonghi vacuum cleaner and a Swift compact water sprayer. All the sampling equipment were thoroughly washed with deionized water prior to sampling to avoid cross contamination. For build-up sampling, dry vacuuming was initially undertaken three times in a perpendicular direction (see Fig. 2.3). Subsequently, the road surface was dampened

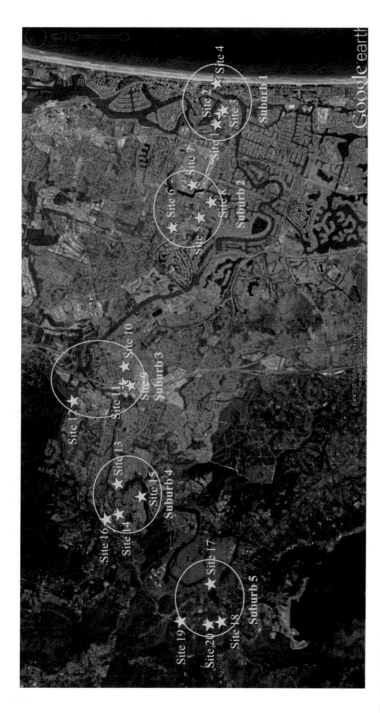

Fig. 2.1 Study sites

Table 2.1 Traffic and land use characteristics of the study sites

Suburbs	Study sites	Land use	Land use characteristics	Daily traffic volume (DTV)	Traffic characteristics	Reference image of the study site
Suburb 1	Site 1	Commercial and residential	Residential houses, offices, school and church	750	Relatively low traffic volume	
	Site 2	Commercial and residential	Residential houses, offices, school and church	750	Relatively low traffic volume	
	Site 3	Commercial	Tourism related area with holiday apartments, shops, restaurants	3000	Parking bays, loading zones and traffic congestion	
	Site 4	Commercial	Shops, school, hospital, offices and residential houses	3000	Collector road with relatively high traffic volume and numerous traffic lights	
Suburb 2	Site 5	Commercial and residential	Residential houses, connected to an arterial road, school, clinic, fuel station, shops, shopping centre and hospital	500	Relatively low traffic volume	

(continued)

Table 2.1 (continued)

Suburbs	Study sites	Land use	Land use characteristics	Daily traffic volume (DTV)	Traffic characteristics	Reference image of the study site
	Site 6	Commercial and residential	Residential houses, school and loading zone	750	Relatively low traffic volume most of the day and high traffic volume prior to and after school time	
	Site 7	Commercial	Numerous motor vehicle businesses	3000	Relatively high traffic volume, parking bays along the roadside and car park	
	Site 8	Commercial and residential	Residential houses, connected to an arterial road, school, clinic, fuel station, shops, shopping centre and hospital	750	Traffic lights, car park, relatively high traffic volume	
Suburb 3	Site 9	Industrial	Large number of industrial premises with a range of industrial activities	3500	Congested traffic	
	Site 10	Industrial	Large number of industrial premises with a range of industrial activities	7000	Congested traffic	

(continued)

Table 2.1 (continued)

Suburbs	Study sites	Land use	Land use characteristics	Daily traffic volume (DTV)	Traffic characteristics	Reference image of the study site
	Site 11	Industrial	Large number of industrial premises with a range of industrial activities	750	Congested traffic	
	Site 12	Industrial	Large number of industrial premises with a range of industrial activities	3500	Congested traffic	
Suburb 4	Site 13	Residential	Large fraction of residential houses	500	Relatively low traffic volume	
	Site 14	Residential	Large fraction of residential houses	750	Relatively low traffic volume	
	Site 15	Residential	Large fraction of residential houses	750	Relatively low traffic volume	

(continued)

Table 2.1 (continued)

Suburbs	Study sites	Land use	Land use characteristics	Daily traffic volume (DTV)	Traffic characteristics	Reference image of the study site
	Site 16	Residential	Large fraction of residential houses	3000	Relatively low traffic volume	
Suburb 5	Site 17	Natural	Very limited anthropogenic activities	750	Relatively low traffic volume	
	Site 18	Natural	Very limited anthropogenic activities	1000	Relatively low traffic volume	
	Site 19	Natural	Very limited anthropogenic activities	150	Relatively low traffic volume	
	Site 20	Natural	Very limited anthropogenic activities	150	Relatively low traffic volume	

Fig. 2.2 A typical sampling plot of 1.0 m × half width of the road (Ma, 2016)

by spraying deionized water and then wet vacuuming was conducted to ensure the collection of any remaining particles present on the road surface. Samples collected by the vacuum cleaner were finally transferred to pre-washed polyethylene bottles and preserved below 4 °C under refrigeration.

2.4 Sample Testing

Build-up samples were separated into two particle size fractions, < 150 and > 150 μm, and both fractions were tested. The separation of samples into two fractions is due to the differences in the build-up process of these two particle size fractions (Jayarathne et al., 2019; Wijesiri et al., 2016). The loads of HMs and PAHs, which have been identified as the most toxic chemical pollutants present in urban stormwater (Ma et al. 2017a, 2017b, 2021) were determined for the two particle size fractions. The investigated HMs are given in Chap. 1. The investigated PAHs comprised of 15 of the 16 priority PAHs species identified by the US EPA (USEPA, 1984) as listed in Chap. 1. Benzo[k]flouranthene was not tested because it was not present in the standards used. Total solids (TS) in the build-up samples in the size fractions < 150 and > 150 μm

Fig. 2.3 Collection of build-up sample using dry vacuuming (Ma, 2016)

were also tested as it is a critical parameter which influences the concentrations and risks posed by HMs and PAHs (Gnecco et al., 2019). Build-up samples were digested according to Standard Method 3030E (APHA, 2017) and tested using an Agilent 8800 Triple Quadrupole Inductively Coupled Plasma Mass Spectrometer (ICP-MS) according to US EPA Method 200.8 (USEPA, 1994) to determine the HM concentrations present. Aqueous build-up samples were digested using liquid–liquid extraction according to US EPA Method 610 (USEPA, 1984), while particulate build-up samples were digested using the accelerated solvent extraction (ASE) procedure method according to US EPA Method 3545 (USEPA, 1996) for testing for PAH concentrations. The concentrations of PAHs were determined using a Shimadzu Gas Chromatograph Mass Spectrometer (GC–MS) TQ8030.

2.5   Summary

This chapter presents the details of the selected study site characteristics, build-up sample collection and sample testing procedures adopted. From a total of five suburbs in the Gold Coast region in Australia, 20 road sites were selected. These sites represented varying traffic and land use characteristics in order to investigate the influence of these characteristics on the risk posed by HMs and PAHs in the context of stormwater reuse. Build-up samples from the identified road surfaces at the selected study sites were collected using a dry and wet vacuuming method. Sample testing focused on nine HMs and 15 PAHs as the primary toxic chemical pollutants present in the collected build-up samples. Additionally, the TS concentrations in the < 150 and > 150 μm size fractions were also determined as this is a critical parameter influencing the concentrations and risks posed by HMs and PAHs present in urban stormwater.

References

APHA. (2017). *Standard methods for the examination of water and wastewater* (23rd ed.). Washington DC.

Egodawatta, P., Thomas, E., & Goonetilleke, A. (2007). Mathematical interpretation of pollutant wash-off from urban road surfaces using simulated rainfall. *Water Research, 41*(13), 3025–3031.

Gnecco, I., Palla, A., & Sansalone, J. J. (2019). Partitioning of zinc, copper and lead in urban drainage from paved source area catchments. *Journal of Hydrology, 578*, 124128.

Gunawardana, C., Egodawatta, P., & Goonetilleke, A. (2014). Role of particle size and composition in metal adsorption by solids deposited on urban road surfaces. *Environmental Pollution, 184*, 44–53.

Gunawardena, J., Ziyath, A. M., Egodawatta, P., Ayoko, G. A., & Goonetilleke, A. (2014). Mathematical relationships for metal build-up on urban road surfaces based on traffic and land use characteristics. *Chemosphere, 99*, 267–271.

Jayarathne, A., Egodawatta, P., Ayoko, G. A., & Goonetilleke, A. (2018). Assessment of ecological and human health risks of metals in urban road dust based on geochemical fractionation and potential bioavailability. *Science of the Total Environment, 635*, 1609–1619.

Jayarathne, A., Egodawatta, P., Ayoko, G. A., & Goonetilleke, A. (2019). Transformation processes of metals associated with urban road dust: A critical review. *Critical Reviews in Environmental Science and Technology, 49*(18), 1675–1699.

Ma, Y. (2016). *Human health risk of toxic chemical pollutants generated from traffic and land use activities.* Queensland University of Technology.

Ma, Y., Liu, A., Egodawatta, P., McGree, J., & Goonetilleke, A. (2017a). Assessment and management of human health risk from toxic metals and polycyclic aromatic hydrocarbons in urban stormwater arising from anthropogenic activities and traffic congestion. *Science of the Total Environment, 579*, 202–211.

Ma, Y., McGree, J., Liu, A., Deilami, K., Egodawatta, P., & Goonetilleke, A. (2017b). Catchment scale assessment of risk posed by traffic generated heavy metals and polycyclic aromatic hydrocarbons. *Ecotoxicology and Environmental Safety, 144*, 593–600.

Ma, Y., Mummullage, S., Wijesiri, B., Egodawatta, P., McGree, J., Ayoko, G. A., & Goonetilleke, A. (2021). Source quantification and risk assessment as a foundation for risk management of metals in urban road deposited solids. *Journal of Hazardous Materials, 408*, 124912.

USEPA. (1984). Guidelines establishing test procedures for the analysis of pollutants under clean water act: Method 610 – polynuclear aromatic hydrocarbons. *Washington, DC, US Environmental Protection Agency 49*, 43344–43352.

USEPA. (1994). Metod 200.8: Trace elements in waters and wastes by inductively coupled plasma-mass spectrometry. *Washington, DC, US Environmental Protection Agency.*

USEPA. (1996). Pressurized fluid extraction (PFE). *Washington, DC, US Environmental Protection Agency.*

Wijesiri, B., Egodawatta, P., McGree, J., & Goonetilleke, A. (2016). Assessing uncertainty in pollutant build-up and wash-off processes. *Environmental Pollution, 212*, 48–56.

Chapter 3
Linking Traffic and Land Use to Stormwater Quality

Abstract Safe reuse of urban stormwater requires risk assessment and management, for which accurate quality estimation is essential. This chapter initially discusses how pollutants build-up data obtained from the field was converted to concentrations in stormwater runoff through build-up and wash-off replication. Mathematical equations are provided to link pollutant concentrations in stormwater to traffic volume and land use area percentages. These equations can be employed as an urban planning tool for stormwater management with changing traffic and land use characteristics without having to undertake resource intensive field monitoring. The research outcomes provide practical recommendations for urban stormwater quality assessment and management.

Keywords Stormwater quality · Stormwater pollutant processes · Pollutant concentration estimation · Traffic volume · Land use · Mathematical modelling

3.1 Background

Toxic chemical pollutants in urban stormwater can pose risk to human health (Hong et al., 2018). Heavy metals (HMs) and polycyclic aromatic hydrocarbons (PAHs) in urban stormwater are the most common toxic pollutants contributed by traffic and land use activities that pose risk to human health (Jayarathne et al., 2018; Ma et al., 2019). Stormwater risk is dependent on the pollutant concentrations present. Therefore, risk assessment requires accurate estimation of HM and PAH concentrations in stormwater. Concentrations of HMs and PAHs in stormwater runoff are primarily influenced by their build-up load on urban impervious surfaces prior to a rainfall event (Wijesiri et al., 2016). Therefore, prediction of the concentrations of HMs and PAHs in stormwater based on their build-up loads provides an essential foundation for the subsequent risk assessment. Urban stormwater quality is also impacted by various anthropogenic activities common to urban areas (Mummullage et al., 2016; Perera et al., 2021). Since traffic and land use related activities play a significant role in influencing urban stormwater quality (Gunawardena et al., 2014; Miranda et al.,

© The Author(s), under exclusive license to Springer Nature Singapore Pte Ltd. 2023   23
Y. Ma et al., *Human Health Risk Assessment of Toxic Chemical Pollutants in Stormwater*,
SpringerBriefs in Water Science and Technology,
https://doi.org/10.1007/978-981-19-9616-0_3

2022), linking the concentrations of HMs and PAHs present in stormwater to traffic and land use characteristics is essential for effective stormwater risk management.

This chapter presents approaches for the prediction of stormwater quality and for mathematically linking stormwater quality to traffic and land use. Accordingly, this chapter initially discusses the prediction of HM and PAH concentrations in urban stormwater through pollutant build-up and wash-off replication and hydraulic and hydrologic modeling. Primary data used as inputs was obtained from the pollutant build-up sampling and testing discussed in Chap. 2. Based on the predicted concentrations of HMs and PAHs in stormwater, a series of mathematical equations were developed to link HM and PAH concentrations to traffic and land use characteristics, which provided the foundation for the assessment of the quantitative risk posed by these pollutants.

3.2 Predicting HM and PAH Concentrations in Stormwater Runoff

This section presents the methodology adopted for predicting HM and PAH concentrations in urban stormwater. As HMs and PAHs are primarily associated with solids in stormwater (Miranda et al., 2021), their concentrations can be determined based on the total solids (TS) concentration present. Accordingly, the methodology adopted entailed build-up and wash-off predictions to estimate TS loads originating from impervious surfaces and the application of hydrologic modelling tools to estimate runoff volumes.

3.2.1 Estimation of Total Solids (TS) Load in Stormwater

Stormwater TS load estimation consisted of three steps; build-up replication, wash-off replication, and impervious area calculation. Build-up and wash-off replication was aimed at estimating build-up and wash-off load from a unit area of an impervious surface. As road and roof surfaces are the two main impervious surface types present in urban areas, the impervious area calculation aimed at determining the area of road and roof surfaces at the study sites.

3.2.1.1 Build-Up Replication

Build-up replication was undertaken on roads and roofs separately due to the differences in build-up coefficients for the two surface types. Pollutant build-up is a dynamic process undergoing changes at all times. Sartor et al. (1974) developed a power equation to replicate the build-up process of solids present on roads.

Egodawatta (2007) further refined this equation to allow consideration of the pre-existing build-up amount after the preceding rainfall event. If the initial amount of build-up is less than the two-day equivalent build-up load (B_{twoday}), build-up load on roads is estimated according to Eq. 3.1 which indicates that the build-up rate is high during the first two dry days. In this context, build-up will vary linearly from the initial amount to B_{twoday} during the first two antecedent dry days. After two days, build-up concentration will vary according to the power equation and will reach the maximum value after about 21 days. After 21 days, build-up load keeps a constant value equal to the maximum load. If the initial build-up amount was greater than B_{twoday}, build-up variation will be linear from the initial amount to the maximum load (Ma, 2016). It is noteworthy that Eq. 3.1 is adopted for estimating the build-up load on road surfaces.

$$B = \begin{cases} linear\ increase\ from\ initial\ amount\ to\ B_{twoday} & D \leq 2 \\ aD^b & 2 < D \leq 21 \\ a \times 21^b & D > 21 \end{cases} \quad (3.1)$$

where:

B TS build-up load on a road (g/m^2);
B_{twoday} two-day equivalent build-up load (g/m^2);
D antecedent dry days ($D > 2$ and the pre-existing build-up amount is less than B_{twoday});
a, b coefficients.

Simulation of the build-up process for a given site requires site specific build-up coefficients a and b. The coefficient b is dependent on the road surface type and a constant value of 0.16 was used for bitumen road surfaces as recommended by Egodawatta (2007). The coefficient a was determined according to the actual build-up load derived from the field study discussed in Chap. 2. Furthermore, rainfall data from the closest rainfall stations to the study sites were used to determine the rainfall intensity for each rainfall event and the number of corresponding dry days for each field sampling episode. In this regard, data from Gold Coast Seaway rainfall station was used for the eight sites in Suburb 1 and Suburb 2. Rainfall data recorded at Hinze Dam rainfall station was adopted for the 12 study sites representing Suburb 3, Suburb 4 and Suburb 5. Sample collection was conducted in the year 2014. It was assumed that the build-up would reach its maximum load over a 21 day dry period (Egodawatta, 2007). Therefore, a long dry period of 24 antecedent dry days prior to the rainfall event on 22 July was selected based on the rainfall record for 2014 for the Gold Coast Seaway in order to determine the coefficient a at the study sites in Suburb 1 and Suburb 2 for build-up replication (Sect. 3.2.1.1) and wash-off replication (Sect. 3.2.1.2). Rainfall records at the Hinze Dam for 2014 showed that the rainfall event on 28 March was very high which washed off nearly 100% of the pollutants built-up according to the wash-off replication (Sect. 3.2.1.2). Accordingly, the build-up load after this rainfall event was assumed to be zero so that the coefficient a at the study sites in Suburb 3, Suburb 4 and Suburb 5 could be determined by

Table 3.1 Coefficient *a* and *b* for road surfaces at the study sites

Study sites	*a*	*b*
Site 1	4.35	0.16
Site 2	1.89	
Site 3	8.26	
Site 4	1.27	
Site 5	4.20	
Site 6	2.44	
Site 7	5.56	
Site 8	0.97	
Site 9	1.56	
Site 10	10.41	
Site 11	7.89	
Site 12	5.51	
Site 13	1.57	
Site 14	2.54	
Site 15	2.38	
Site 16	5.69	
Site 17	1.32	
Site 18	0.47	
Site 19	3.53	
Site 20	3.90	

replicating the build-up (Sect. 3.2.1.1) and wash-off (Sect. 3.2.1.2) from the 28 March rainfall event to the sampling date. Estimation results of coefficients *a* and *b* at the study sites are listed in Table 3.1.

The build-up replication procedure discussed above was applied to the road surfaces. Pollutant build-up on roof surfaces is different from road surfaces. Egodawatta (2007) found that build-up load on roads and roofs at the same site are mathematically related and can be calculated according to Eq. 3.2.

$$B_{Roof} = \begin{cases} \frac{0.26}{0.92} \times B_{Road} \\ \frac{0.43 \times D^{0.266}}{1.65 \times D^{0.16}} \times B_{Road} \end{cases} \tag{3.2}$$

Where

B_{Roof} and B_{Road} total solids build-up load on roof and road surfaces (g/m^2).

3.2.1.2 Wash-Off Replication

Wash-off replication relates to the fraction of build-up solids transported by stormwater runoff. Sartor et al. (1974) developed an exponential equation to replicate wash-off based on the assumption that the pollutants deposited during the dry period are completely washed-off by the subsequent rainfall event (Eq. 3.3). Egodawatta (2007) further refined the wash-off equation based on the hypothesis that only a fraction of the pollutants are washed-off during a rainfall event. Accordingly, a new parameter C_F, named as capacity factor, was introduced into the exponential equation as shown in Eq. 3.4. According to Egodawatta (2007), the value of C_F is dependent on the surface type and rainfall intensity as depicted in Eqs. 3.5 and 3.6.

$$F_W = \frac{W}{W_0} = 1 - e^{-KIt} \tag{3.3}$$

$$F_W = C_F\left(1 - e^{-KIt}\right) \tag{3.4}$$

where

F_W fraction wash-off;
C_F capacity factor;
W weight of solids washed-off after time t (g/m^2);
W_0 initial weight of solids on the road surface (g/m^2);
K wash-off coefficient (constant 8×10^{-4} for road surfaces and 9.33×10^{-3} for roof surfaces);
I rainfall intensity (mm/h);
t rainfall duration (h).

$$C_{F_{Road}} = \begin{cases} (0.01 \times I) + 0.1 & 5 \sim 40 \text{ mm/h} \\ 0.5 & 40 \sim 90 \text{ mm/h} \\ (0.0098 \times I) - 0.38 & 90 \sim 133 \text{ mm/h} \end{cases} \tag{3.5}$$

where

$C_{F_{Road}}$ capacity factor for road surfaces;
I rainfall intensity (mm/h).

$$C_{F_{Roof}} = \begin{cases} (0.008 \times I) + 0.59 & 5 \sim 40 \text{ mm/h} \\ 0.91 & 40 \sim 90 \text{ mm/h} \\ (0.0036 \times I) + 0.59 & 90 \sim 133 \text{ mm/h} \end{cases} \tag{3.6}$$

where

$C_{F_{Roof}}$ capacity factor for roof surfaces;
I rainfall intensity (mm/h).

3.2.1.3 Impervious Area Calculation

Egodawatta (2007) recommended that build-up and wash-off processes should be replicated using a small plot to estimate stormwater quality in order to ensure surface homogeneity. Accordingly, this study selected a unit catchment at each study site to estimate the pollutant load. The unit catchment was a rectangular area where the sampling site was located in the middle. The rectangular catchment covered the entire urban surfaces beside the road to a length of 100 m and an appropriate width. It was assumed that stormwater risk was mainly posed by HMs and PAHs in the runoff from the impervious surfaces. This is due to the high impervious area percentage and runoff coefficient for the urban area. The impervious area comprised of roads (including driveways) and roofs. It was assumed that pollutants build-up load on driveways was the same as the sampled roads due to the sources and build-up process. Roads (including driveways) and roofs were demarcated using Google Earth and their area was determined using ArcMap Geographical information systems (GIS) software. Figure 3.1 gives an example of the unit catchment and impervious area demarcation for Site 16. The impervious surface area for all the study sites are given in Table 3.2.

3.2.2 Stormwater Runoff Volume Estimation

Stormwater runoff volume was estimated using hydraulic and hydrologic modelling based on representative rainfall events. Simulation was based on individual rainfall events rather than continuous rainfall data in order to build an adequate database for

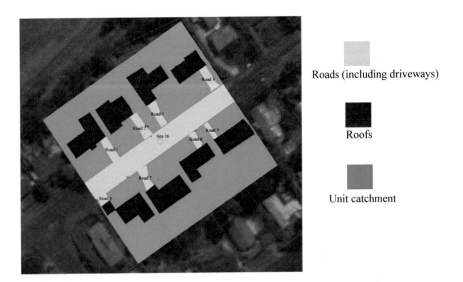

Fig. 3.1 Unit catchment and impervious surface area at Site 16 (Ma, 2016)

Table 3.2 Impervious area at the study sites

Study sites	Road (including driveway) area (m^2)	Roof area (m^2)
Site 1	2995	3027
Site 2	1578	2773
Site 3	1961	1963
Site 4	1898	2481
Site 5	1550	3072
Site 6	1732	1107
Site 7	3706	7856
Site 8	1781	2809
Site 9	4040	3863
Site 10	2798	8740
Site 11	2624	1475
Site 12	3164	4507
Site 13	1129	1998
Site 14	1433	2239
Site 15	827	844
Site 16	1637	2047
Site 17	1076	0
Site 18	1132	0
Site 19	576	0
Site 20	1120	0

the subsequent mathematical simulation of pollutant concentrations. MIKE URBAN software was selected for simulating stormwater runoff volume.

3.2.2.1 Representative Rainfall Data Selection

Pollutant concentration in stormwater is influenced by rainfall characteristics (Liu et al., 2013). In order to accurately estimate pollutant concentration, appropriate rainfall data was required. Rainfall data during the last ten years (i.e. 2004–2013) before the sampling year 2014 recorded by the two selected rainfall stations were obtained. Annual rainfall depth from 2004 to 2013 were compared and the rainfall data recorded in year 2008 was selected due to its annual rainfall depth being the closest to the average over the ten years (Fig. 3.2). Consequently, 57 rainfall events from the Gold Coast Seaway and 55 rainfall events from Hinze Dam rainfall stations were selected.

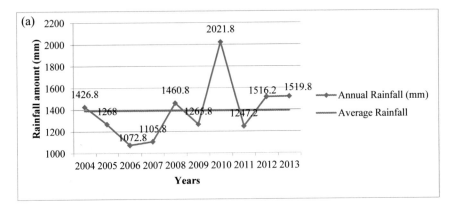

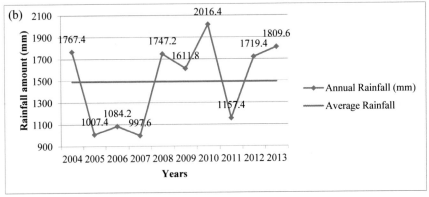

Fig. 3.2 Annual rainfall depths **a** Gold Coast Seaway; and **b** Hinze Dam over the period 2004–2013 (Ma, 2016)

3.2.2.2 Runoff Volume Simulation for the Selected Rainfall Events

The simulation procedure consisted of four steps; input of catchment data, definition of rainfall time series, input of network data, and simulation of catchment and network. Catchment data required the input of the area of the sub-catchments in the unit catchment defined in Sect. 3.2.1.3. Three sub-catchments, namely, roads (including driveways), roofs and pervious surfaces, were justified based on their difference in initial loss. Time series defined in this study was rainfall depth per minute (mm/min) for the selected rainfall events during the representative year 2008. Network defined that the runoff generated from the sub-catchments drained together to a node and discharged from the outlet of the unit catchment (Fig. 3.3). Catchment and network simulation estimated the total runoff volume generated from the unit catchment at each study site.

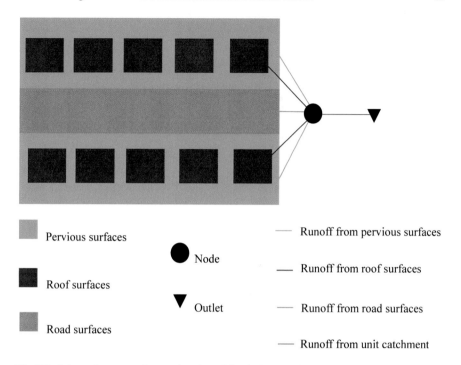

Pervious surfaces ⬤ Node — Runoff from pervious surfaces

— Runoff from roof surfaces

Roof surfaces ▼ Outlet — Runoff from road surfaces

Road surfaces

— Runoff from unit catchment

Fig. 3.3 Sub-catchments and networks adopted for the hydrologic model (Ma, 2016)

3.2.3 Pollutant Concentrations in Stormwater Runoff

Estimation of pollutant concentrations in stormwater provided the foundation for the risk assessment (Ma et al., 2017). HM and PAH concentrations were estimated based on TS concentration in stormwater derived according to the methodology discussed in Sects. 3.2.1 and 3.2.2. This was based on the hypothesis that adsorption fraction of HMs and PAHs to solids in build-up and wash-off were the same.

Since HMs and PAHs are mainly adsorbed to solids in stormwater, it was evident that the HMs and PAHs associated with TS could pose the highest risk to human health. A relatively higher fraction of HMs and PAHs are adsorbed to fine solids (solids < 150 μm) (Gunawardana et al., 2014). Moreover, fine solids have higher mobility in stormwater and as such, exert a relatively higher risk. Therefore, it was assumed that the risk from HMs and PAHs associated with TS and fine (< 150 μm) solids, respectively, represented the maximum and minimum risk level posed by stormwater. Accordingly, concentrations of HMs and PAHs adsorbed to both, TS and fine (< 150 μm) solids in stormwater were estimated. The estimation was conducted using Eq. 3.7.

$$C = EMC \times B \times 10^{-6} \qquad (3.7)$$

where

C concentration of chemical pollutants attached to TS or fine ($< 150 \, \mu$m) solids in stormwater runoff (mg/L);

EMC Concentration of TS or fine ($< 150 \, \mu$m) solids in stormwater runoff (mg/L);

B average concentration of HMs and PAHs in build-up solids (mg/kg).

3.3   Linking Traffic and Land Use to Pollutant Concentrations

Pollutant concentrations in urban stormwater is impacted by traffic and land use factors (Gunawardena et al., 2014). Therefore, it was hypothesised that HM and PAH concentrations can be estimated as a function of the traffic volume and different land use area percentages as shown in Eq. 3.8. Traffic volume data and land use area percentages at the study sites are given in Table 3.3. It is noteworthy that the sum of commercial, industrial and residential land use area percentages was less than 100%. This was because the pervious area was not demarcated in assessing the urban land use area. Due to the high impervious area ratio in the urban areas and the large runoff coefficient of pervious surfaces, it was assumed that urban stormwater risk arises from the toxic effects of pollutants in runoff from impervious surfaces.

$$C = f(DTV, C, I, R) \qquad (3.8)$$

where

C HM and PAH concentration in stormwater (mg/L);

DTV daily traffic volume (vehicles per day (VPD));

C, I, R percentage of commercial, industrial and residential land use area within 1 km radius from the investigated location.

Equation 3.8 was modelled via a linear geostatistical model with consideration of spatial variability in the concentration data (Ma et al., 2016). Logarithmic transformation of pollutant concentrations (ln C) and traffic volume (ln DTV) was carried out to ensure the linear relationship between dependent and independent variables. Accordingly, the expected ln C was modelled using Eq. 3.9.

$$E[lnC|DTV, C, I \text{ and } R] = \beta_0 + \beta_1 \times f(\ln DTV) + \beta_2$$
$$\times f(C) + \beta_3 \times f(I) + \beta_4 \times f(R) \qquad (3.9)$$

where

Table 3.3 Daily traffic volume and land use area percentages at the study sites

Study sites	Daily traffic volume (VPD)	Commercial area percentage (%)	Industrial area percentage (%)	Residential area percentage (%)
Site 1	750	11	0	45
Site 2	750	18	0	40
Site 3	3000	29	0	21
Site 4	3000	16	0	42
Site 5	500	19	0	52
Site 6	750	18	0	58
Site 7	3000	50	0	34
Site 8	750	29	0	51
Site 9	3500	14	20	31
Site 10	7000	6	17	22
Site 11	750	14	19	31
Site 12	3500	19	2	24
Site 13	500	0	0	62
Site 14	750	0	0	32
Site 15	750	0	0	63
Site 16	3000	0	0	34
Site 17	750	0	0	0
Site 18	1000	0	0	0
Site 19	150	0	0	0
Site 20	150	0	0	0

Note Traffic volume data were provided by Gold Coast City Council

$E[\ln C|DTV, C, I \text{ and } R]$	the expected value of logarithm of pollutants concentration in stormwater (ln (mg/L));
$\beta_0, \beta_1, \beta_2, \beta_3 \text{ and } \beta_4,$	the coefficients of the independent variables;
$\ln DTV$	the logarithm of daily traffic volume (ln(VPD));
$C, I \text{ and } R$	the percentage of commercial, industrial and residential land use areas (%);
$f(\ln DTV), f(C), f(I) \text{ and } f(R)$	a function of $\ln DTV, C, I$ and R.

To capture any spatial variability in the data, Eq. 3.9 was modelled using geoR package in R software which provides a conducive environment for geostatistical computing and graphics (Diggle & Ribeiro, 2007). The Matern function in geoR package was selected as the covariance model. There was no trend evident for the residuals variability which confirmed that the model fitted the data well. Figure 3.4 shows an example of residuals variability estimated for benzo[a]pyrene (B[a]P) adsorbed to TS. The line y = x was employed to assess whether the estimated values agree well with the observed data. Figure 3.5 shows the distribution of the

estimated B[a]P concentrations and observed data. According to the modelling procedure discussed above, concentrations of HMs and PAHs associated with both, TS and
fine (< 150 μm) solids in stormwater were assessed. Tables 3.4, 3.5, 3.6 and 3.7
present the summary of the assessment results.

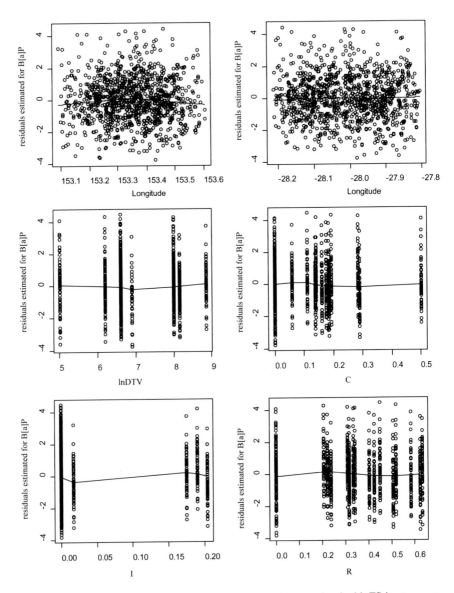

Fig. 3.4 Estimated residual variability of B[a]P concentration associated with TS in stormwater
(trend represented by the line in the plot) (Ma, 2016)

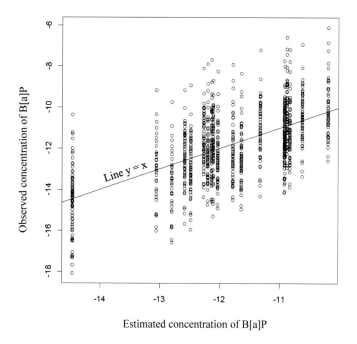

Fig. 3.5 Distribution of estimated values and observed data for B[a]P associated with TS in stormwater (Ma, 2016)

HM concentrations were determined using the equations in Table 3.4. Figure 3.6 illustrates HM concentrations in stormwater at each site and the comparison with water quality guidelines for recreation (ANZECC, 2000) and consumption (NHMRC & AWRC, 2011). The blue lines represent the allowable threshold values of HMs in water for recreational use and the red lines refer to the allowable threshold concentrations of HMs in potable water. Some figures do not show the threshold values as these values are either above the highest HM concentrations at the study sites or the water quality guidelines do not provide the threshold concentrations. It can be noted that Al and Fe concentrations from all the sites were over the threshold concentrations in water for recreational use. Pb and Mn concentrations at some sites were higher than the threshold values. These results indicate that stormwater in the study area cannot be used for recreational activities such as swimming. In addition, Pb and Ni concentrations at some sites exceed the threshold concentrations for drinking water, suggesting that stormwater from these sites cannot be reused for consumption purpose without appropriate treatment.

Figure 3.7 shows the average concentrations of HMs in stormwater runoff generated from the four roads within each suburb. Generally, HM concentrations were the highest for industrial areas, followed by commercial and residential areas. Concentration of HMs in stormwater from the natural areas was the lowest.

Concentrations of PAHs in stormwater at each site were estimated according to equations in Table 3.5. Figure 3.8 presents the estimated PAH concentrations at the

Table 3.4 Results for concentration assessment of HMs in TS (Ma, 2016)

Equations
$Al = \exp(5.058 - 2.047 \times \ln DTV + 0.1743 \times (\ln DTV)^2 + 1.731 \times C + 1.092 \times R)$
$Cr = \exp(-8.6244 + 0.2630 \times \ln DTV + 2.9025 \times C + 25.1244 \times I - 100.2798 \times I^2 + 0.8461 \times R)$
$Mn = \exp(-3.851 + 2.466 \times C + 45.08 \times I - 212.2 \times I^2 + 0.5679 \times R)$
$Fe = \exp(-1.3398 + 0.2629 \times \ln DTV + 2.5012 \times C + 21.7752 \times I - 96.2173 \times I^2)$
$Ni = \exp(-8.7869 + 0.2321 \times \ln DTV + 4.0288 \times C + 54.0466 \times I - 203.4114 \times I^2 + 0.5166 \times R)$
$Cu = \exp(-8.653 + 0.4779 \times \ln DTV + 5.841 \times C + 24.78 \times I - 93.96 \times I^2 + 0.557 \times R)$
$Zn = \exp(-6.8975 + 0.4189 \times \ln DTV + 3.1094 \times C + 46.8731 \times I - 213.1362 \times I^2 + 1.2630 \times R)$
$Cd = \exp(-14.62 + 0.6534 \times \ln DTV + 2.014 \times C - 6.845 \times C^2 + 4.056 \times I - 1.290 \times R)$
$Pb = \exp(-8.2979 + 0.4144 \times \ln DTV + 3.1837 \times C + 25.0152 \times I - 98.0013 \times I^2)$

DTV: daily traffic volume (vehicles per day (VPD)); C, I, R: percentage of commercial, industrial and residential land use area within 1 km radius from interested location; exp/ln: exponential/logarithmic function

study sites. ANZECC (2000) and NHMRC and AWRC (2011) provide threshold concentrations for only B[a]P in water for recreational and potable use, while threshold values for other PAHs are not provided by the water quality guidelines. The maximum acceptable concentration of B[a]P is 0.00001 mg/L and the concentration of B[a]P at many urban sites exceeds this value. This suggests that PAHs in stormwater at these sites would pose cancer risk to humans.

The average concentration of PAHs in stormwater runoff from the four roads in each suburb is shown in Fig. 3.9. Compared to HM concentrations in stormwater, the estimated PAH concentrations showed less variability among the various land uses. PAH concentrations were also the highest in the stormwater generated from the industrial area and the lowest in the stormwater from the natural area. PAH concentrations in stormwater from commercial and residential areas were similar.

3.4 Summary

This chapter discussed the translation of pollutants build-up load to the concentration present in stormwater runoff. Furthermore, a series of mathematical equations were developed to predict HM and PAH concentrations in stormwater based on daily traffic

Table 3.5 Results for concentration assessment of PAHs in TS (Ma, 2016)

Equations
NAP = exp(−13.0057 + 2.2244 × C + 44.0004 × I-208.5743 × I^2 + 2.3271 × R)
ACY = exp(−18.7764 + 0.6287 × lnDTV-3.8347 × C + 11.3567 × C^2 + 22.2219 × I-98.2006 × I^2 + 2.7677 × R)
ACE = exp(−16.8329 + 0.3191 × lnDTV-3.4542 × C + 7.6039 × C^2 + 2.7041 × I + 1.2821 × R)
FLU = exp(−15.0477 + 0.2792 × lnDTV-5.2849 × C + 12.1146 × C^2 + 25.0081 × I-105.4251 × I^2 + 1.2495 × R)
PHE = exp(−12.7752 + 0.2335 × lnDTV-1.8874 × C + 6.3417 × C^2 + 23.2440 × I-99.5676 × I^2 + 1.4360 × R)
ANT = exp(−15.7314 + 0.4039 × lnDTV + 3.3674 × I)
FLA = exp(−13.9610 + 0.3248 × lnDTV-8.9481 × C + 22.6337 × C^2 + 47.6350 × I-209.3615 × I^2 + 0.9829 × R)
PYR = exp(−13.83 + 0.4109 × lnDTV-3.378 × C + 10.16 × C^2 + 23.32 × I-105.7 × I^2 + 1.961 × R)
BaA = exp(−19.1925 + 0.8233 × lnDTV + 1.3639 × C + 1.4070 × I + 1.6338 × R)
CHR = exp(−16.72 + 0.6626 × lnDTV + 1.976 × C + 2.293 × I + 0.9365 × R)
BbF = exp(−17.67 + 0.7247 × lnDTV + 1.903 × C + 1.216 × R)
BaP = exp(−18.87 + 0.8781 × lnDTV + 2.277 × C + 1.491 × R)
IND = exp(−17.03 + 0.6054 × lnDTV-3.352 × C + 13.33 × C^2 + 30.97 × I-144.9 × I^2 + 1.762 × R)
DaA = exp(−17.0912 + 0.5192 × lnDTV + 2.2610 × C + 25.0942 × I-116.0339 × I^2 + 1.6652 × R)
BgP = exp(−15.42 + 0.5412 × lnDTV + 36.87 × I-169.1 × I^2 + 2.414 × R)

DTV: daily traffic volume (vehicles per day (VPD)); C, I, R: percentage of commercial, industrial and residential land use area within 1 km radius from interested location; exp/ln: exponential/logarithmic function

volume and urban land use area percentages. These equations can be employed for estimating stormwater pollutant concentrations without complex field monitoring. These equations can also be used for the formulation of stormwater management strategies. Although the equations derived in this research project are site specific, they present an innovative approach to stormwater quality estimation and management. The modelling results showed the HM and PAH concentrations in stormwater varied with different land uses. Stormwater pollution was the most serious in the industrial areas, followed by commercial and residential areas, while stormwater

Table 3.6 Results for concentration assessment of HMs in solids < 150 μm (Ma, 2016)

Equations
$Al = \exp(-13.3078 + 2.6629 \times \ln DTV - 0.1532 \times (\ln DTV)^2 - 4.5435 \times C + 12.4244 \times C^2 + 44.2164 \times I - 204.9952 \times I^2 + 1.3108 \times R)$
$Cr = \exp(-26.9111 + 4.8640 \times \ln DTV - 0.3112 \times (\ln DTV)^2 - 2.5523 \times C + 12.1061 \times C^2 + 47.9003 \times I - 204.0476 \times I^2 + 0.5839 \times R)$
$Mn = \exp(-22.8705 + 4.4254 \times \ln DTV - 0.2716 \times (\ln DTV)^2 - 4.9288 \times C + 12.5600 \times C^2 + 55.7913 \times I - 259.5832 \times I^2 + 1.0352 \times R)$
$Fe = \exp(-13.6836 + 2.9047 \times \ln DTV - 0.1676 \times (\ln DTV)^2 + 1.0934 \times C + 44.7496 \times I - 206.2285 \times I^2 + 0.7611 \times R)$
$Ni = \exp(-27.9284 + 5.1500 \times \ln DTV - 0.3345 \times (\ln DTV)^2 + 2.7383 \times C + 72.2912 \times R - 285.8606 \times R^2)$
$Cu = \exp(-22.7836 + 3.9198 \times \ln DTV - 0.2264 \times (\ln DTV)^2 + 4.4768 \times C + 29.4201 \times I - 115.5464 \times I^2 + 0.5945 \times R)$
$Zn = \exp(-17.4434 + 2.7467 \times \ln DTV - 0.1437 \times (\ln DTV)^2 + 1.8575 \times C + 42.1926 \times I - 188.1349 \times I^2 + 1.6716 \times R)$
$Cd = \exp(-24.1876 + 2.8568 \times \ln DTV - 0.1491 \times (\ln DTV)^2 - 1.2810 \times C)$
$Pb = \exp(-11.3228 + 0.6123 \times \ln DTV - 2.3663 \times C + 11.9864 \times C^2 + 34.7190 \times I - 140.0913 \times I^2 + 0.9244 \times R)$

DTV: daily traffic volume (vehicles per day (VPD)); C, I, R: percentage of commercial, industrial and residential land use area within 1 km radius from interested location; exp/ln: exponential/logarithmic function

quality in the natural areas was the best. Approaches for stormwater reuse for recreational and potable purposes are discussed in water quality guidelines. Based on the results obtained, stormwater from all the study sites were not suitable for recreational use such as swimming and stormwater from some of the study sites cannot be reused for consumption.

Table 3.7 Results for concentration assessment of PAHs in solids < 150 μm (Ma, 2016)

Equations
NAP = $\exp(-16.3586 + 0.3389 \times \ln DTV\text{-}3.6059 \times C + 10.9032 \times C^2 + 26.2000 \times I\text{-}114.5088 \times I^2 + 2.7176 \times R)$
ACY = $\exp(-33.9416 + 4.5322 \times \ln DTV\text{-}0.2551 \times (\ln DTV)^2\text{-}7.2402 \times C + 15.8078 \times C^2 + 3.6249 \times I + 2.0391 \times R)$
ACE = $\exp(-33.4097 + 4.6431 \times \ln DTV\text{-}0.2887 \times (\ln DTV)^2\text{-}7.6697 \times C + 13.2273 \times C^2 + 3.1553 \times I + 0.7180 \times R)$
FLU = $\exp(-24.5371 + 2.4227 \times \ln DTV\text{-}0.1281 \times (\ln DTV)^2\text{-}9.4743 \times C + 18.6256 \times C^2 + 5.7119 \times I + 0.7102 \times R)$
PHE = $\exp(-15.7154 + 0.4913 \times \ln DTV + 2.9205 \times I + 1.3899 \times R)$
ANT = $\exp(-23.0036 + 1.9973 \times \ln DTV\text{-}0.0982 \times (\ln DTV)^2\text{-}1.1322 \times C + 3.1344 \times I)$
FLA = $\exp(-17.1510 + 0.6483 \times \ln DTV\text{-}11.0649 \times C + 24.0136 \times C^2 + 6.6748 \times I + 1.1306 \times R)$
PYR = $\exp(-16.7735 + 0.6869 \times \ln DTV\text{-}5.2783 \times C + 11.9617 \times C^2 + 2.5159 \times I + 2.1876 \times R)$
BaA = $\exp(-33.4407 + 4.4578 \times \ln DTV\text{-}0.2404 \times (\ln DTV)^2\text{-}3.8908 \times C + 9.5423 \times C^2 + 3.0062 \times I + 1.4018 \times R)$
CHR = $\exp(-25.441 + 2.593 \times \ln DTV\text{-}0.118 \times (\ln DTV)^2\text{-}1.810 \times C + 6.126 \times C^2 + 3.071 \times I + 1.506 \times R)$
BbF = $\exp(-30.9075 + 4.0341 \times \ln DTV\text{-}0.2181 \times (\ln DTV)^2\text{-}2.5665 \times C + 7.6256 \times C^2 + 2.5254 \times I + 1.4022 \times R)$
BaP = $\exp(-29.7000 + 3.4408 \times \ln DTV\text{-}0.1608 \times (\ln DTV)^2 + 0.8037 \times C + 1.4459 \times R)$
IND = $\exp(-27.2442 + 2.8800 \times \ln DTV\text{-}0.1307 \times (\ln DTV)^2\text{-}6.2993 \times C + 15.8776 \times C^2 + 2.4545 \times I + 1.7137 \times R)$
DaA = $\exp(-35.1046 + 5.2444 \times \ln DTV\text{-}0.3172 \times (\ln DTV)\text{-}4.3010 \times C + 12.7056 \times C^2 + 31.1761 \times I\text{-}137.4866 \times I^2 + 1.6662 \times R)$
BgP = $\exp(-25.9924 + 3.0257 \times \ln DTV\text{-}0.1545 \times (\ln DTV)^2\text{-}2.5337 \times C + 29.9913 \times I\text{-}131.4012 \times I^2 + 2.4693 \times R)$

DTV: daily traffic volume (vehicles per day (VPD)); C, I, R: percentage of commercial, industrial and residential land use area within 1 km radius from interested location; exp/ln: exponential/logarithmic function

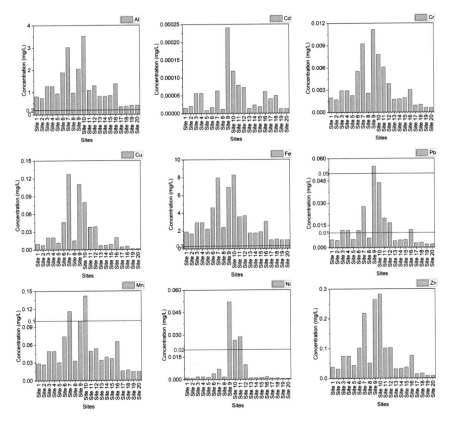

Fig. 3.6 Estimated HM concentrations in stormwater at each study site (*Note* Blue and red lines represent the allowable threshold values for HMs for recreational and potable use, respectively)

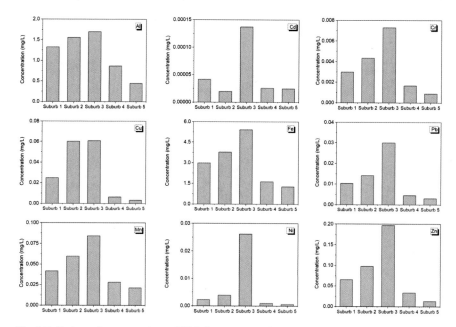

Fig. 3.7 Estimated concentrations of HMs in stormwater in each suburb

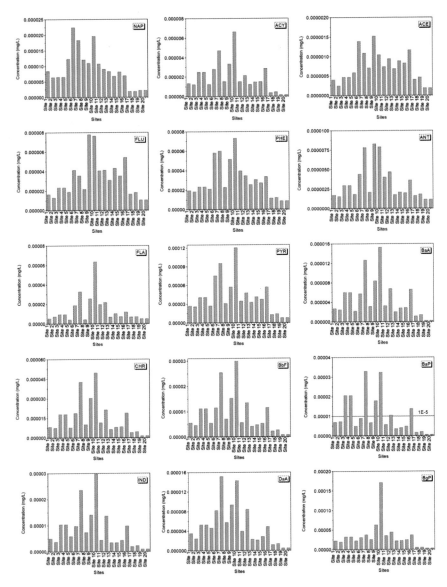

Fig. 3.8 Estimated PAH concentrations in stormwater at each study site (*Note* Blue line represents the allowable threshold values for B[a]P in water for recreational and potable use)

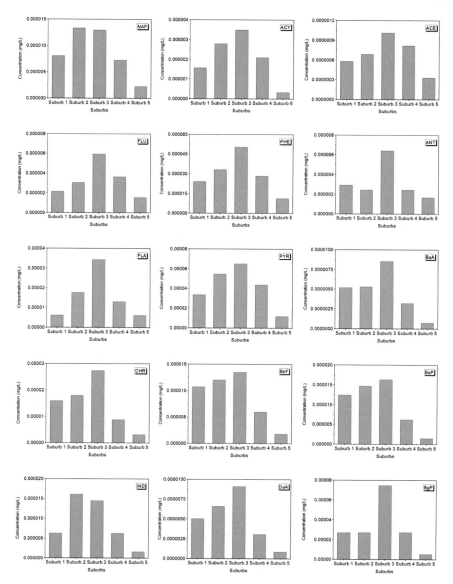

Fig. 3.9 Estimated concentrations of PAHs in stormwater in each suburb

References

ANZECC. (2000). Australian and New Zealand guidelines for fresh and marine water quality. Canberra, Australian and New Zealand Environment and Conservation Council and Agriculture and Resource Management Council of Australia and New Zealand, pp. 1–103.

Diggle, P. & Ribeiro, P. J. (2007). *Model-based geostatistics*. Springer Science & Business Media.

Egodawatta, P. (2007). *Translation of small-plot scale pollutant build-up and wash-off measurements to urban catchment scale.* Queensland University of Technology.

Gunawardana, C., Egodawatta, P., & Goonetilleke, A. (2014). Role of particle size and composition in metal adsorption by solids deposited on urban road surfaces. *Environmental Pollution, 184,* 44–53.

Gunawardena, J., Ziyath, A. M., Egodawatta, P., Ayoko, G. A., & Goonetilleke, A. (2014). Mathematical relationships for metal build-up on urban road surfaces based on traffic and land use characteristics. *Chemosphere, 99,* 267–271.

Hong, N., Liu, A., Zhu, P., Zhao, X., Guan, Y., Yang, M., & Wang, H. (2018). Modelling benzene series pollutants (BTEX) build-up loads on urban roads and their human health risks: Implications for stormwater reuse safety. *Ecotoxicology and Environmental Safety , 164*(NOV.), 234–242.

Jayarathne, A., Egodawatta, P., Ayoko, G. A., & Goonetilleke, A. (2018). Assessment of ecological and human health risks of metals in urban road dust based on geochemical fractionation and potential bioavailability. *Science of the Total Environment, 635,* 1609–1619.

Liu, A., Egodawatta, P., Guan, Y., & Goonetilleke, A. (2013). Influence of rainfall and catchment characteristics on urban stormwater quality. *Science of the Total Environment, 444,* 255–262.

Ma, Y. (2016). *Human health risk of toxic chemical pollutants generated from traffic and land use activities.* Queensland University of Technology.

Ma, Y., Deilami, K., Egodawatta, P., Liu, A., McGree, J., & Goonetilleke, A. (2019). Creating a hierarchy of hazard control for urban stormwater management. *Environmental Pollution, 255,* 113217.

Ma, Y., Egodawatta, P., McGree, J., Liu, A., & Goonetilleke, A. (2016). Human health risk assessment of heavy metals in urban stormwater. Science of the Total Environment.

Ma, Y., McGree, J., Liu, A., Deilami, K., Egodawatta, P., & Goonetilleke, A. (2017). Catchment scale assessment of risk posed by traffic generated heavy metals and polycyclic aromatic hydrocarbons. *Ecotoxicology and Environmental Safety, 144,* 593–600.

Miranda, L. S., Deilami, K., Ayoko, G. A., Egodawatta, P., & Goonetilleke, A. (2022). Influence of land use class and configuration on water-sediment partitioning of heavy metals. *Science of the Total Environment, 804,* 150116.

Miranda, L. S., Wijesiri, B., Ayoko, G. A., Egodawatta, P., & Goonetilleke, A. (2021). Water-sediment interactions and mobility of heavy metals in aquatic environments. *Water Research, 202,* 117386.

Mummullage, S., Egodawatta, P., Ayoko, G. A., & Goonetilleke, A. (2016). Use of physicochemical signatures to assess the sources of metals in urban road dust. *Science of the Total Environment, 541*(3), 1303.

NHMRC&AWRC. (2011). *Australian drinking water guidelines.* Australian Government Publishing Service.

Perera, T., McGree, J., Egodawatta, P., Jinadasa, K. B. S. N., & Goonetilleke, A. (2021). A Bayesian approach to model the trends and variability in urban stormwater quality associated with catchment and hydrologic parameters. *Water Research, 197,* 117076.

Sartor, J. D., Boyd, G. B., & Agardy, F. J. (1974). Water pollution aspects of street surface contaminants. *Journal Water Pollution Control Federation, 46*(3), 458–467.

Wijesiri, B., Egodawatta, P., McGree, J., & Goonetilleke, A. (2016). Influence of uncertainty inherent to heavy metal build-up and wash-off on stormwater quality. *Water Research, 91,* 264–276.

Chapter 4
Human Health Risk Assessment for Toxic Chemical Pollutants in Urban Stormwater

Abstract Accurate and efficient risk assessment of toxic chemical pollutants present in urban stormwater is essential for safely utilizing this potential water resource. The mathematical equations developed in Chap. 3 were applied for predicting stormwater quality using quantitative parameters including traffic volume and land use area percentages. This chapter firstly presents a detailed risk assessment approach based on the quantitative parameters. However, quantitative parameters alone are not adequate to interpret the associated risk. As such, the influence of qualitative parameters (i.e. traffic and land use characteristics) in relation to stormwater risk was also analyzed. Consequently, two qualitative parameters, including fraction of vehicle related businesses (FVS) and braking and starting frequency (BSF) were selected to further refine the risk model developed. The eventual risk assessment model is expected to provide practical approaches for stormwater risk management in the context of stormwater reuse and urban planning.

Keywords Risk assessment · Stormwater risk · Heavy metals · Polycyclic aromatic hydrocarbons · Traffic characteristics · Land use characteristics · Stormwater quality · Stormwater pollutant processes

4.1 Background

Urban stormwater reuse is becoming increasingly popular to overcome the serious water scarcity being experienced around the world (Ma et al., 2019). However, pollutants such as heavy metals (HMs) and polycyclic aromatic hydrocarbons (PAHs) which are present in stormwater can exert adverse impacts on human health (Fabian et al., 2022). Therefore, in order to enable utilizing the full potential offered by stormwater resources, efficient and accurate assessment of the human health risk posed by HMs and PAHs is critical. As discussed in Chap. 3, traffic and land use are the two most important factors influencing stormwater quality. Accordingly, stormwater risk could be evaluated based on traffic and land use related parameters.

This chapter initially discusses the development of a risk assessment model based on mathematical equations presented in Chap. 3 for estimating HM and PAH

© The Author(s), under exclusive license to Springer Nature Singapore Pte Ltd. 2023   45
Y. Ma et al., *Human Health Risk Assessment of Toxic Chemical Pollutants in Stormwater*,
SpringerBriefs in Water Science and Technology,
https://doi.org/10.1007/978-981-19-9616-0_4

concentrations in stormwater runoff. Impact of traffic and land use characteristics on stormwater risk is then discussed and appropriate qualitative parameters were selected to refine the initial model. The outcomes presented are expected to contribute to robust decision making in relation to urban stormwater management and reuse.

4.2 Risk Assessment Model Development

Generally, the risk assessment procedure contains four steps: hazard identification, exposure assessment, dose–response assessment and risk characterization as discussed in Chap. 1. Figures 4.1 and 4.2 present the recommended risk assessment procedure for the risk posed by HMs and PAHs in urban stormwater, respectively.

4.2.1 Hazard Identification

The main concern in relation to HMs is their non-carcinogenic effects such as hypertension, renal dysfunction, Parkinson's disease, allergic reactions and skin ulcers (Ma et al., 2017a, 2017b, 2017c). In terms of non-carcinogenic effects, a threshold value is generally given to determine whether the chemical substance can exert toxic effects on human health as discussed in Chap. 1. Therefore, the intake of HMs below the threshold value is considered safe, and if the intake exceeds the threshold value, HMs would exert toxic effects on people. Different from HMs, the carcinogenic effect of PAHs on human health is of significant concern (Ma et al., 2017a, 2017b, 2017c). In terms of carcinogenic effects, the probability of cancer risk posed by the toxic chemical substances is usually considered as discussed in Chap. 1. Once the probability is higher than the threshold value, PAHs would be considered to pose cancer risk. In addition to the toxicity of HMs and PAHs on human health, their ubiquitous nature and persistence in stormwater are also of concern (Patel et al., 2018).

4.2.2 Exposure Assessment

People can come into contact with HMs and PAHs in stormwater through three exposure pathways, namely, ingestion of stormwater by drinking, accidental ingestion of stormwater while swimming and dermal contact with stormwater (Ma et al., 2017a, 2017b, 2017c). Exposure through each pathway can be quantified using Eqs. 4.1, 4.2 and 4.3 proposed by the United States Environmental Protection Agency (USEPA, 1989).

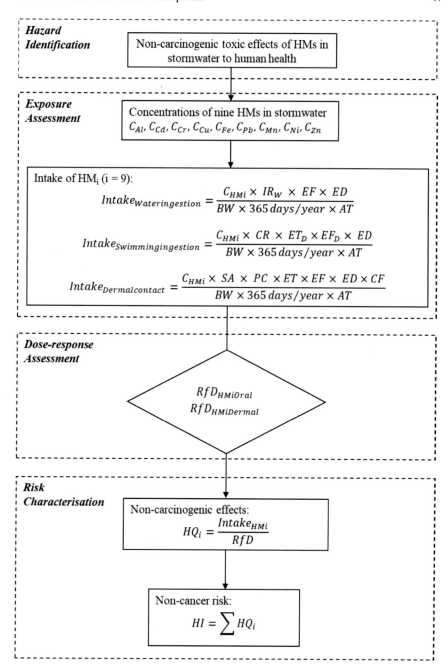

Fig. 4.1 Calculation procedure for determining the non-carcinogenic risk posed by HMs (*Note* Parameters are defined in Sects. 4.2.1–4.2.4) (Ma, 2016)

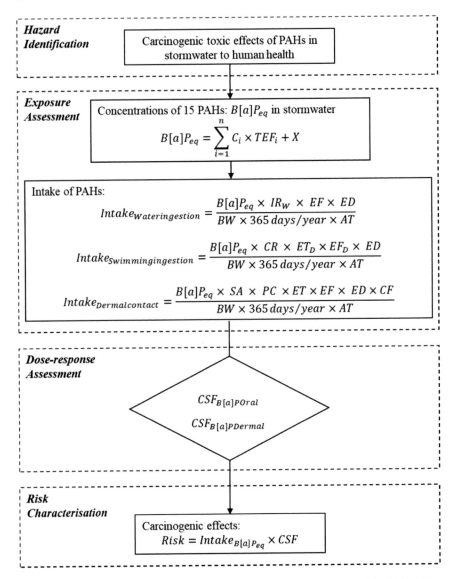

Fig. 4.2 Calculation procedure for determining the carcinogenic risk posed by PAHs (*Note* parameters are defined in Sects. 4.2.1–4.2.4) (Ma, 2016)

$$Intake_{Wateringestion} = \frac{C \times IR_W \times EF \times ED}{BW \times 365 \text{ days/year} \times AT} \tag{4.1}$$

where

$Intake_{Wateringestion}$ intake by ingestion of stormwater (mg/(kg day));
C concentration of HMs or PAHs in stormwater (mg/L);

IR_W	water ingestion rate (L/day);
EF	exposure frequency (days/year);
ED	exposure duration (years);
BW	body weight of an adult (kg);
AT	average time of exposure (years).

$$Intake_{Swimmingingestion} = \frac{C \times CR \times ET_D \times EF_D \times ED}{BW \times 365 days/year \times AT} \quad (4.2)$$

where

$Intake_{Swimmingingestion}$	intake by accidental ingestion (mg/(kg day));
C	concentration of HMs or PAHs in stormwater (mg/L);
CR	contact rate (L/h);
ET_D	exposure time (hours/event);
EF_D	exposure frequency (events/year);
ED	exposure duration (years);
BW	body weight (kg);
AT	average time of exposure (years).

$$Intake_{Dermalcontact} = \frac{C \times SA \times PC \times ET_D \times EF_D \times ED \times CF}{BW \times 365 \, days/year \times AT} \quad (4.3)$$

where

$Intake_{DermalContact}$	intake by dermal contact (mg/(kg day));
C	concentration of HMs or PAHs in stormwater (mg/L);
SA	skin surface area available for contact (cm^2);
PC	chemical-specific dermal permeability constant (cm/hour);
ET_D	exposure time (hours/event);
EF_D	exposure frequency (events/year);
ED	exposure duration (years);
CF	volumetric conversion for water;
BW	body weight (kg);
AT	average time during exposure (years).

Parameter values in Eqs. 4.1–4.3 were justified according to professional judgement by USEPA (1989) based on the assumption that people are exposed to HMs and PAHs in stormwater with the maximum amount through a specific pathway. Table 4.1 summarizes values of the parameters recommended by USEPA (1989).

HM and PAH concentrations in stormwater were estimated using the mathematical equations presented in Chap. 3. Exposure to HMs was assessed based on individual HMs because the risk from different HMs can be combined (Ma et al., 2016). Concentration of each HM derived as discussed in Chap. 3 was directly substituted

Table 4.1 Parameter values for exposure assessment (USEPA, 1989)

Parameters	IR_W (L/day)	EF (days/year)	ED (years)	BW (kg)	AT (years)	CR (L/h)	ET_D (h/event)	EF_D (events/year)	SA (cm^2)	PC (cm/h)	CF (L/cm^3)
Values	2	365	30	70	70	0.05	2.6	7	18,000	0.7	1/1000

Table 4.2 TEF values for the 15 priority PAHs (Adapted from Nisbet and LaGoy (1992))

Compounds	TEFs
Naphthalene	0.001
Acenaphthylene	0.001
Acenaphthene	0.001
Anthracene	0.01
Phenanthrene	0.001
Fluorene	0.001
Fluoranthene	0.001
Pyrene	0.001
Benzo[a]anthracene	0.1
Chrysene	0.01
Benzo[a]pyrene	1
Benzo[b]fluoranthene	0.1
Benzo[ghi]perylene	0.01
Indeno[1,2,3-cd]pyrene	0.1
Dibenz[a,h]anthracene	5

into Eqs. 4.1–4.3. In contrast, exposure to PAHs was assessed as a group and the toxic equivalent factor (TEF) approach was adopted (Gbeddy et al., 2020). TEF approach selects benzo[a]pyrene (B[a]P) as the reference and assigns TEFs (Table 4.2) relative to B[a]P for the other PAHs. Accordingly, concentration of the PAH mixture was determined as the B[a]P-equivalent concentration (B[a]P$_{eq}$) as shown in Eq. 4.4.

$$B[a]P_{eq} = \sum_{i=1}^{n} C_i \times TEF_i + X \qquad (4.4)$$

where

$B[a]P_{eq}$ the B[a]P equivalent exposure presented by the mixture;
C_i concentration of the ith PAH in the mixture;
TEF_i toxic equivalent factor of the ith PAH in the mixture;
n number of PAHs in the mixture;
X concentration of B[a]P in the mixture.

4.2.3 Dose–Response Assessment

As discussed in Chap. 1, a reference dose (RfD) defined by USEPA (1989) is adopted for assessing non-carcinogenic effects, while a cancer slope factor (CSF) is adopted for assessing carcinogenic effects. According to Hazard Identification, the main concern in relation to the presence of HMs and PAHs in stormwater is non-carcinogenic and carcinogenic effects on human health, respectively. Therefore, the RfD of HMs and CSF of PAHs need to be determined. Exposure Assessment identifies three pathways involving oral and dermal contact (USEPA, 1989). Therefore, RfD of each HM and CSF of B[a]P through oral (RfD_{oral} and CSF_{oral}) and dermal (RfD_{dermal} and CSF_{dermal}) exposure were determined according to the Integrated Risk Information System (IRIS, 2015) and are given in Table 4.3.

4.2.4   Risk Characterization

4.2.4.1   HMs Risk Characterization

Potential non-carcinogenic effects of HMs in stormwater were estimated using the hazard quotient (HQ) as discussed in Chap. 1 and hazard index (HI) methods. HQ represents the effects of an individual HM on human health (Eq. 4.5), while HI assesses the combined toxic effects of multiple HMs on human health and is evaluated as the summation of HQ for the individual HMs (Eq. 4.6). Either HQ or HI less than 1 indicates that HMs in stormwater do not pose a risk to human health, while HQ or HI greater than 1 means that HMs would exert a risk to human health.

$$HQ = \frac{Intake}{RfD} \tag{4.5}$$

where

HQ hazard quotient for a given HM through oral or dermal contact;
$Intake$ daily intake of a given HM in stormwater through oral or dermal contact (mg/(kg day));
RfD reference dose of a given HM through oral or dermal contact (mg/(kg day)).

$$HI = \sum HQ_{ioral} + \sum HQ_{idermal} \tag{4.6}$$

where

HI hazard index for multiple HMs through oral and dermal contact.

4.2.4.2   PAHs Risk Characterization

Cancer risk posed by PAHs in stormwater was estimated as incremental lifetime cancer risk (ILCR) as discussed in Chap. 1 (Eq. 4.7). ILCR below 10^{-6} means that people would not suffer from adverse health impacts from PAHs in stormwater according to USEPA (1989). ILCR ranging from 10^{-6} to 10^{-4} indicates that PAHs

Table 4.3 RfDs of HMs and CSF of B[a]P (IRIS, 2015)

Pollutants	RfD_{oral} (mg/kg day)$^{-1}$	RfD_{dermal}(mg/kg day)$^{-1}$	CSF_{oral}(mg/kg day)$^{-1}$	CSF_{dermal}(mg/kg day)$^{-1}$
Al	1	0.1	–	–
Cr	0.003	0.00006	–	–
Mn	0.02	0.00184	–	–
Fe	0.7	–	–	–
Ni	0.02	0.0054	–	–
Cu	0.04	0.012	–	–
Zn	0.3	0.06	–	–
Cd	0.0005	0.00001	–	–
Pb	0.0035	0.000525	–	–
B[a]P	–	–	7.3	25

are potentially carcinogenic to human health while the probability is still acceptable. ILCR above 10^{-4} means that PAHs have a significant probability to pose cancer risk.

$$ILCR = Intake \times CSF \tag{4.7}$$

where

$ILCR$ incremental lifetime cancer risk;
$Intake$ daily intake of PAHs in stormwater (mg/(kg day));
CSF cancer slope factor of B[a]P $((mg/(kg\ day))^{-1})$.

4.2.5   Risk Assessment Model

Risk posed by HMs and PAHs in stormwater to human health was estimated according to the four steps of risk assessment as outlined in Figs. 4.1 and 4.2. As shown, stormwater risk is dependent on HM and PAH concentrations in stormwater. Since pollutant concentrations were based on daily traffic volume (DTV) and land use area percentages at the study sites as discussed in Chap. 3, stormwater risk can also be estimated based on DTV and land use area percentages. Besides, Chap. 3 noted that the concentrations of HMs and PAHs associated with total and fine solids (solids < 150 μm) in stormwater can be applied for assessing the maximum and minimum level of risk. Accordingly, risk models were developed to estimate human health risks from HMs and PAHs based on DTV and land use area percentages shown as Eqs. 4.8–4.11 (Ma, 2016). The values of DTV and land use area percentages at the study sites can be found in Table 3.3 in Chap. 3.

$$
\begin{aligned}
Risk_{HMtotal} = {} & 0.05 \times e^{5.058-2.047\times lnDTV+0.1743\times (lnDTV)^2+1.731\times C+1.092\times R} \\
& + 68.20 \times e^{-8.6244+0.2630\times lnDTV+2.9025\times C+25.1244\times I-100.2798\times I^2+0.8461\times R} \\
& + 2.70 \times e^{-3.851+2.466\times C+45.08\times I-212.2\times I^2+0.5679\times R} \\
& + 0.02 \times e^{-1.3398+0.2629\times lnDTV+2.5012\times C+21.7752\times I-96.2173\times I^2} \\
& + 1.33 \times e^{-8.7869+0.2321\times ln\,DTV+4.0288\times C+54.0466\times I-203.4114\times I^2+0.5166\times R} \\
& + 0.63 \times e^{-8.653+0.4779\times lnDTV+5.841\times C+24.78\times I-93.96\times I^2+0.557\times R} \\
& + 0.10 \times e^{-6.8975+0.4189\times lnDTV+3.1094\times C+46.8731\times I-213.1362\times I^2+1.2630\times R} \\
& + 409.18 \times e^{-14.62+0.6534\times lnDTV+2.014\times C-6.845\times C^2+4.056\times I-1.290\times R} \\
& + 10.83 \times e^{-8.2979+0.4144\times lnDTV+3.1837\times C+25.0152\times I-98.0013\times I^2}
\end{aligned}
\tag{4.8}
$$

$$
\begin{aligned}
Risk_{HMfine} = {} & 0.05 \times e^{-13.3078+2.6629\times lnDTV-0.1532\times (lnDTV)^2-4.5435\times C+12.4244\times C^2+44.2164\times I-204.9952\times I^2+1.3108\times R} \\
& + 68.20 \times e^{-26.9111+4.8640\times lnDTV-0.3112\times (lnDTV)^2-2.5523\times C+12.1061\times C^2+47.9003\times I-204.0476\times I^2+0.5839\times R} \\
& + 2.70 \times e^{-22.8705+4.4254\times lnDTV-0.2716\times (lnDTV)^2-4.9288\times C+12.5600\times C^2+55.7913\times I-259.5832\times I^2+1.0352\times R} \\
& + 0.02 \times e^{-13.6836+2.9047\times lnDTV-0.1676\times (lnDTV)^2+1.0934\times C+44.7496\times I-206.2285\times I^2+0.7611\times R} \\
& + 1.33 \times e^{-27.9284+5.1500\times lnDTV-0.3345\times (lnDTV)^2+2.7383\times C+72.2912\times R-285.8606\times R^2} \\
& + 0.63 \times e^{-22.7836+3.9198\times lnDTV-0.2264\times (lnDTV)^2+4.4768\times C+29.4201\times I-115.5464\times I^2+0.5945\times R}
\end{aligned}
$$

$+\,0.10 \times e^{-17.4434+2.7467\times\ln\text{DTV}-0.1437\times(\ln\text{DTV})^2+1.8575\times C+42.1926\times I-188.1349\times I^2+1.6716\times R}$

$+\,409.18 \times e^{-24.1876+2.8568\times\ln\text{DTV}-0.1491\times(\ln\text{DTV})^2-1.2810\times C}$

$+\,10.83 \;\times e^{-11.3228+0.6123\times\ln\text{DTV}-2.3663\times C+11.9864\times C^2+34.7190\times I-140.0913\times I^2+0.9244\times R}$ $\hspace{2em}$ (4.9)

$$\text{Risk}_{\text{PAHtotal}} = \frac{332,059.3}{1,788,500}$$

$\times\,0.001 \times e^{-13.0057+2.2244\times C+44.0004\times I-208.5743\times I^2+2.3271\times R}$

$+\,0.001 \times e^{-18.7764+0.6287\times\ln\text{DTV}-3.8347\times C+11.3567\times C^2+22.2219\times I-98.2006\times I^2+2.7677\times R}$

$+\,0.001 \times e^{-16.8329+0.3191\times\ln\text{DTV}-3.4542\times C+7.6039\times C^2+2.7041\times I+1.2821\times R}$

$+\,0.001 \times e^{-15.0477+0.2792\times\ln\text{DTV}-5.2849\times C+12.1146\times C^2+25.0081\times I-105.4251\times I^2+1.2495\times R}$

$+\,0.001 \times e^{-12.7752+0.2335\times\ln\text{DTV}-1.8874\times C+6.3417\times C^2+23.2440\times I-99.5676\times I^2+1.4360\times R}$

$+\,0.01 \times e^{-15.7314+0.4039\times\ln\text{DTV}+3.3674\times I}$

$+\,0.001 \times e^{-13.9610+0.3248\times\ln\text{DTV}-8.9481\times C+22.6337\times C^2+47.6350\times I-209.3615\times I^2+0.9829\times R}$

$+\,0.001 \times e^{-13.83+0.4109\times\ln\text{DTV}-3.378\times C+10.16\times C^2+23.32\times I-105.7\times I^2+1.961\times R}$

$+\,0.1 \times e^{-19.1925+0.8233\times\ln\text{DTV}+1.3639\times C+1.4070\times I+1.6338\times R}$

$+\,0.01 \times e^{-16.72+0.6626\times\ln\text{DTV}+1.976\times C+2.293\times I+0.9365\times R}$

$+\,0.1 \times e^{-17.67+0.7247\times\ln\text{DTV}+1.903\times C+1.216\times R}$

$+\,1 \times e^{-18.87+0.8781\times\ln\text{DTV}+2.277\times C+1.491\times R}$

$+\,0.1 \times e^{-17.03+0.6054\times\ln\text{DTV}-3.352\times C+13.33\times C^2+30.97\times I-144.9\times I^2+1.762\times R}$

$+\,5 \times e^{-17.0912+0.5192\times\ln\text{DTV}+2.2610\times C+25.0942\times I-116.0339\times I^2+1.6652\times R}$

$+\,0.01 \times e^{-15.42+0.5412\times\ln\text{DTV}+36.87\times I-169.1\times I^2+2.414\times R}$ $\hspace{2em}$ (4.10)

$$\text{Risk}_{\text{PAHfine}} = \frac{332,059.3}{1,788,500}$$

$\times\,0.001 \times e^{-16.3586+0.3389\times\ln\text{DTV}-3.6059\times C+10.9032\times C^2+26.2000\times I-114.5088\times I^2+2.7176\times R}$

$+\,0.001 \times e^{-33.9416+4.5322\times\ln\text{DTV}-0.2551\times(\ln\text{DTV})^2-7.2402\times C+15.8078\times C^2+3.6249\times I+2.0391\times R}$

$+\,0.001 \times e^{-33.4097+4.6431\times\ln\text{DTV}-0.2887\times(\ln\text{DTV})^2-7.6697\times C+13.2273\times C^2+3.1553\times I+0.7180\times R}$

$+\,0.001 \times e^{-24.5371+2.4227\times\ln\text{DTV}-0.1281\times(\ln\text{DTV})^2-9.4743\times C+18.6256\times C^2+5.7119\times I+0.7102\times R}$

$+\,0.001 \times e^{-15.7154+0.4913\times\ln\text{DTV}+2.9205\times I+1.3899\times R}$

$+\,0.01 \times e^{-23.0036+1.9973\times\ln\text{DTV}-0.0982\times(\ln\text{DTV})^2-1.1322\times C+3.1344\times I}$

$+\,0.001 \times e^{-17.1510+0.6483\times\ln\text{DTV}-11.0649\times C+24.0136\times C^2+6.6748\times I+1.1306\times R}$

$+\,0.001 \times e^{-16.7735+0.6869\times\ln\text{DTV}-5.2783\times C+11.9617\times C^2+2.5159\times I+2.1876\times R}$

$+\,0.1 \times e^{-33.4407+4.4578\times\ln\text{DTV}-0.2404\times(\ln\text{DTV})^2-3.8908\times C+9.5423\times C^2+3.0062\times I+1.4018\times R}$

$+\,0.01 \times e^{-25.441+2.593\times\ln\text{DTV}-0.118\times(\ln\text{DTV})^2-1.810\times C+6.126\times C^2+3.071\times I+1.506\times R}$

$+\,0.1 \times e^{-30.9075+4.0341\times\ln\text{DTV}-0.2181\times(\ln\text{DTV})^2-2.5665\times C+7.6256\times C^2+2.5254\times I+1.4022\times R}$

$+\,1 \times e^{-29.7000+3.4408\times\ln\text{DTV}-0.1608\times(\ln\text{DTV})^2+0.8037\times C+1.4459\times R}$

$+\,0.1 \times e^{-27.2442+2.8800\times\ln\text{DTV}-0.1307\times(\ln\text{DTV})^2-6.2993\times C+15.8776\times C^2+2.4545\times I+1.7137\times R}$

$+\,5 \times e^{-35.1046+5.2444\times\ln\text{DTV}-0.3172\times(\ln\text{DTV})-4.3010\times C+12.7056\times C^2+31.1761\times I-137.4866\times I^2+1.6662\times R}$

$+\,0.01 \times e^{-25.9924+3.0257\times\ln\text{DTV}-0.1545\times(\ln\text{DTV})^2-2.5337\times C+29.9913\times I-131.4012\times I^2+2.4693\times R}$ $\hspace{2em}$ (4.11) where

$Risk_{HMtotal}$, $Risk_{HMfine}$, $Risk_{PAHtotal}$, $Risk_{PAHfine}$	risk posed by HMs and PAHs associated with total and fine solids in stormwater to human health;
DTV	daily traffic volume on the road (vehicles per day);
C, I and R	percentage of commercial, industrial and residential area within 1 km vicinity of the road.

4.3 Risk Posed by Toxic Chemical Pollutants

4.3.1 Risk Posed by HMs

Based on the developed risk assessment model, human health risks posed by HMs at the study sites were estimated. Figures 4.3 and 4.4 shows the human health risks posed by HMs associated with total and fine solids. As evident from Fig. 4.3, HI posed by all HMs associated with total solids (TS) at the five study sites (Site 7, 9, 10, 11 and 12) was > 1, indicating that HMs in stormwater from these sites were toxic to human health. However, HQ of each HM associated with TS from these five sites was < 1, suggesting that individual HM in stormwater was not toxic to human health. These outcomes imply that even though individual HM may be safe to people, the combined effects posed by multiple HMs can be toxic. Therefore, risk assessment and risk management should consider the toxic effects of the multiple HMs in stormwater rather than considering them individually. In addition to the five sites, HI at the other study sites was < 1, suggesting that HMs in stormwater from these sites were not toxic to human health even at the maximum risk level. Figure 4.4 shows that HI of multiple HMs or HQ of a single HM associated with fine solids was < 1 at all the study sites. This indicates that the minimum level of risk posed by HMs in stormwater was safe to human health.

Comparison of the results derived from Figs. 4.3 and 3.6 indicates that stormwater reuse recommendations provided by water quality guidelines (ANZECC, 2000; NHMRC&AWRC, 2011) and the risk assessment undertaken as discussed above can give contradictory results. According to the results derived from Fig. 3.6, HMs in stormwater from all the study sites were toxic to human health for swimming, while the risk assessment procedure developed by this research project indicates that HMs in stormwater from many of the study sites were safe. The difference between Figs. 4.3 and 3.6 confirms that only determining the concentrations of HMs in stormwater cannot provide comprehensive confirmation of safety for reuse. It is recommended that in the case of urban stormwater reuse, it is important to consider the risk from multiple HMs present rather than individual HM concentrations.

The contribution by each HM to the combined risk posed by multiple HMs in stormwater was evaluated and the results are presented in Fig. 4.5. Contribution by Cr to the combined risk was the largest at 34.05%. Mn contributed the second largest

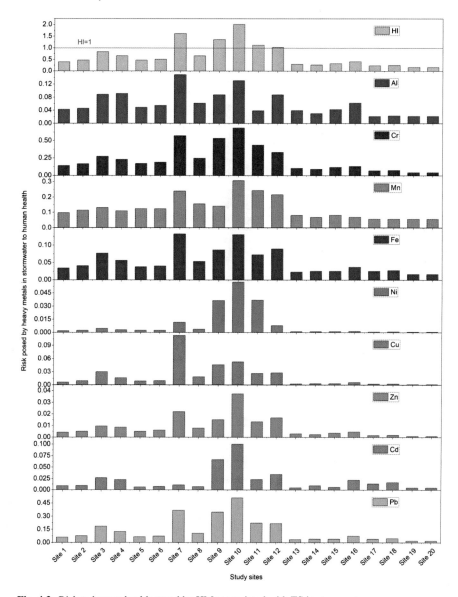

Fig. 4.3 Risk to human health posed by HMs associated with TS in stormwater

at 22.35%, followed by Pb which contributed 17.51% to the combined risk. Contribution by Al and Fe to the combined toxicity was low at 10.34% and 8.61%, while Ni, Cu, Zn and Cd contributed the least at 0.85%, 2.14%, 1.10% and 3.05%, respectively. As discussed in Chap. 3, concentrations of Al and Fe were much higher than other HMs in stormwater, while the contribution by Al and Fe to the combined risk was at moderate level. In contrast, concentrations of Cr, Mn and Pb were relatively

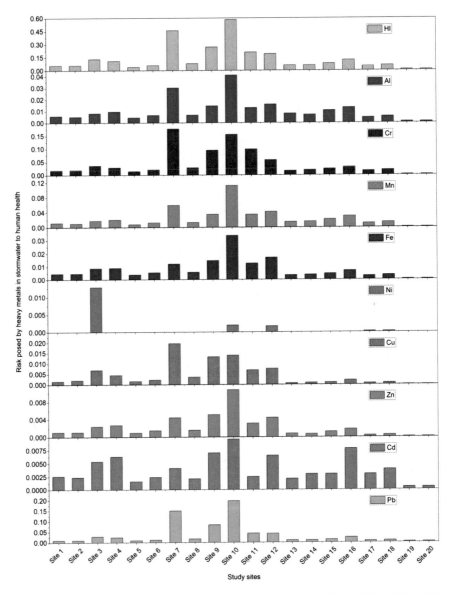

Fig. 4.4 Risk to human health posed by HMs associated with fine solids (solids < 150 μm) in stormwater

low. However, they contributed most to the combined toxicity in the HM mixture, suggesting that these three HMs can exert high toxic effects to human health even at low concentrations. It is further recommended that water quality guidelines should take into consideration HI rather than solely a HM concentration threshold.

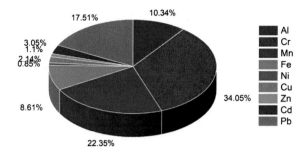

Fig. 4.5 Contribution to the combined risk by individual HMs

4.3.2 Risk Posed by PAHs

Risk to human health posed by PAHs associated with total and fine solids in stormwater was quantified and the results are presented in Figs. 4.6 and 4.7, respectively. The outcomes show that cancer risk posed by PAHs attached to total and fine solids in stormwater from most sites ranged from 10^{-6} to 10^{-4}. This indicates that PAHs in urban stormwater generated from the study area has potential to cause cancer.

Contribution by each PAH to the combined cancer risk posed by the PAH mixture was compared and interpreted as shown in Fig. 4.8. Cancer risk was mainly contributed by PAHs of heavy molecular weight (i.e. PAHs with 5–6 benzene rings)

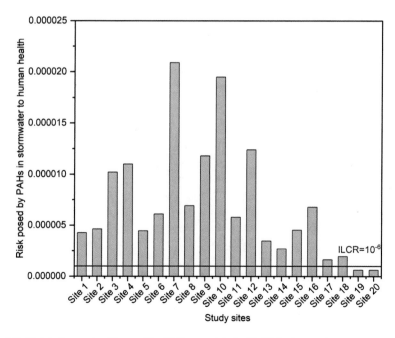

Fig. 4.6 Risk to human health posed by PAHs associated with TS in stormwater

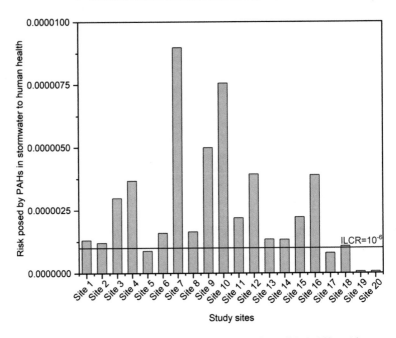

Fig. 4.7 Risk to human health posed by PAHs associated with fine solids (< 150 μm) in stormwater

which accounted for 98% of the combined risk. On the other hand, PAHs of light molecular weight (i.e. PAHs with 2–4 benzene rings) contributed only 2% to the combined risk. This confirms that the risk from the PAH mixture in stormwater is mainly influenced by heavy PAHs.

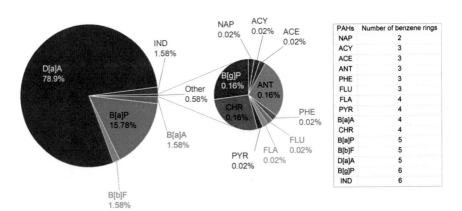

Fig. 4.8 Contribution to the combined risk by individual PAHs

4.3.3 Combined Risk Posed by HMs and PAHs in Stormwater

Based on the developed risk assessment model, the combined human health risk posed by HM and PAH mixtures in stormwater was estimated for the study area in order to identify the priority areas with high risk. Accordingly, a risk map was developed to show the combined stormwater risk generated by the study area (Fig. 4.9). The study area was demarcated by a 1 km × 1 km grid which entailed a total of 96 grid cells. The human health risk arising from HMs and PAHs in stormwater from each grid cell was estimated according to Eqs. 4.8 and 4.10. The land use area percentages within each grid cell were identified using ArcMap software. The road closest to the centroid of a grid cell was identified by ArcMap and the daily traffic volume on this road was used to represent the traffic volume within the grid cell. It is important to note that this approach provides a methodology to identify the traffic volume and its application should consider the spatial distribution of roads within the grid cell. For example, if the closest road to the centroid is a laneway and there is a major arterial road in the cell, it is the traffic data for the arterial road that should be considered. Traffic and land use data for this study was provided by the Gold Coast City Council.

As shown in Fig. 4.9, grid cells in red color mean that HI > 1 and ILCR > 10^{-6}, indicating that both, HMs and PAHs in stormwater from these areas can pose risk to human health. The orange grid cells represent where HI < 1 and ILCR > 10^{-6}, indicating that only PAHs are potentially toxic to human health. The green colored

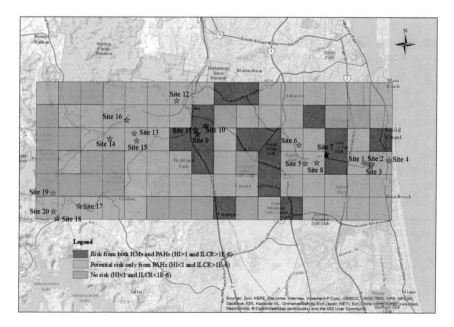

Fig. 4.9 Risk map for combined risk posed by HMs and PAHs in stormwater (Ma et al., 2017a, 2017b, 2017c)

grid cells are where HI < 1 and ILCR < 10^{-6}, implying that stormwater from these areas are safe to people. The risk map prioritizes critical areas which can pose a risk to human health and requires appropriate management measures.

4.4   Influence of Traffic and Land Use on Risk

Traffic volume and land use area percentages are the most important factors influencing risk arising from HMs and PAHs in stormwater. Variation of risk with traffic volume within each land use suburb is presented in Fig. 4.10. Either HI (risk posed by HMs) or ILCR (risk posed by PAHs) generally increased with increasing traffic volume in the same suburb, suggesting that higher traffic volume leads to higher risk posed by HMs and PAHs in stormwater. In addition, risks were relatively higher in stormwater from suburbs with more intense anthropogenic activities (Suburb 1, Suburb 2 and Suburb 3) compared to suburbs with relatively lower anthropogenic activities (Suburb 4) and the natural suburb (Suburb 5). This indicates that industrial and commercial activities result in higher risk from stormwater compared to residential activities in an urban area. Although risk is influenced by daily traffic volume and land use area percentages, risk shows significant variation even with the same traffic volume and land use. For example, Site 7 in Suburb 2, Site 3 and 4 in Suburb 1 are all commercial sites and the traffic volume at the road sites are similar. However, risk from stormwater at Site 7 was much higher than Site 3 and 4. This implies that traffic volume and conventional land use types are not adequate to characterize stormwater risk, while traffic and land use characteristics in the vicinity could also influence the risk related to urban stormwater. Consequently, traffic and land use characteristics in the surrounding area also needs to be considered in risk assessment in relation to HMs and PAHs in stormwater.

Based on the above discussion, human health risk from HMs and PAHs in stormwater should be assessed using daily traffic volume (DTV), land use area percentages (C, I and R), traffic and land use characteristics. Different from DTV, C, I and R, traffic and land use characteristics are not solely expressed in numeric form. For example, the condition of traffic congestion in an urban road and the types of commercial activities at a road site cannot be simply presented as a numeric value. Accordingly, DTV, C, I and R were identified as quantitative variables while traffic and land use characteristics were identified as qualitative variables. In this context, stormwater risk is influenced by both, quantitative and qualitative variables. In order to select the appropriate independent variables, it is necessary to analyze the influence of quantitative and qualitative factors of traffic and land use, to stormwater risk.

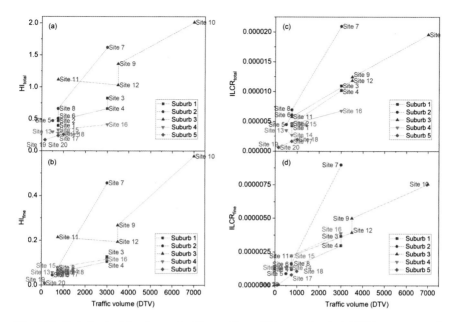

Fig. 4.10 Variation in risk to human health posed by: **a** HMs associated with TS in stormwater; **b** HMs associated with fine solids in stormwater; **c** PAHs associated with TS in stormwater; **d** PAHs associated with fine solids in stormwater

4.4.1 Influence of Quantitative Factors to Stormwater Risk

Influence of daily traffic volume (DTV) and land use area percentage (C, I and R) to human health risk from HMs and PAHs was analyzed using Preference Ranking Organization Method for Enrichment Evaluations (PROMETHEE) and Graphical Analysis for Interactive Assistance (GAIA) methods. PROMETHEE and GAIA methods aim at selecting the most preferred objects taking into consideration the issues of concern. PROMETHEE is adopted to rank objects from the most preferred to the least preferred according to the concerned variables, while GAIA is the graphical representation of the PROMETHEE result. Two data matrices were prepared to represent risk from HMs (HI) and PAHs (ILCR), respectively. Each matrix comprised of 20 rows (the 20 study sites as objects) and five columns (DTV, C, I, R, HI or ILCR as variables). Relationship between risk and traffic volume and land use area percentages was then analyzed using Visual PROMETHEE software and the result is illustrated as GAIA biplots shown in Fig. 4.11. GAIA analysis result shows that either HI or ILCR is positively related to DTV, suggesting that high traffic volume is a primary reason for high stormwater risk. In relation to land use, the angle between the risk index (either HI or ILCR) and I is acute, while the risk index is almost orthogonal to C and R. This indicates that the industrial area exerts a significant influence on stormwater risk, while commercial and residential areas are rarely relevant to stormwater risk. The weak relationship between C and R and risk emphasizes again that conventional land use types cannot comprehensively explain the variation in stormwater risk with land use, and the surrounding land use characteristics should also be considered for estimating human health risk from urban stormwater.

4.4.2 Influence of Qualitative Factors on Stormwater Risk

The qualitative influence of traffic and land use characteristics to stormwater risk was investigated through further detailed field investigations. To enable this, PROMETHEE and GAIA methods were initially used to rank and separate the study sites into different groups based on non-cancer and cancer risk from stormwater. A data matrix with 20 rows (the 20 study sites as objects) and two columns (HI and ILCR as variables) were prepared for the PROMETHEE and GAIA analysis. Table 4.4 gives the PROMETHEE ranking for combined risk posed by HMs and PAHs. Overall, stormwater from Suburb 3 posed the highest human health risk, followed by Suburb 2 and Suburb 1. Stormwater risk from Suburb 4 was generally lower than the other urban suburbs. Risk posed by stormwater from Suburb 5 (natural area) was the lowest.

The graphical representation of Table 4.4 is given in the GAIA biplot shown in Fig. 4.12. As shown in Fig. 4.12, the decision axis Pi points to the right in the biplot, indicating that the stormwater risk increased from sites on the left hand side to the sites on the right hand side. There are seven study sites, namely, Site 3, 4, 7, 9, 10, 11 and 12 which cluster around the positive horizonal axis, suggesting stormwater from these sites posed high risk. In contrast, Site 17, 18, 19 and 20 cluster around the negative horizontal axis and are furthest away from the origin implying that the HMs and PAHs in stormwater from these sites posed low risk. The other nine study sites cluster closer to the origin which indicates that the risk posed by stormwater was moderate. Consequently, the study sites were categorized into three groups as, high risk, moderate risk and low risk. The reasons for the three risk levels are discussed below.

4.4.2.1 High Risk Sites

The high risk posed by stormwater from Site 9, 10, 11 and 12 in Suburb 3 was due to the intensive industrial activities (see Fig. 4.13). These industrial activities were expected to generate significant pollutant loads resulting in high risk from stormwater (Kho et al., 2014). Furthermore, the intense industrial activities would also result in high traffic volume further exacerbating the risk.

Site 7, 3 and 4 are primarily commercial road sites. However, the types of commercial activities at the three sites are different. The main commercial activities at Site 7 are motor vehicle related businesses (Fig. 4.14) contributing significant HMs and PAHs to the road surfaces. Moreover, due to various motor vehicle related services and other commercial business premises, roadside car parking is common at this

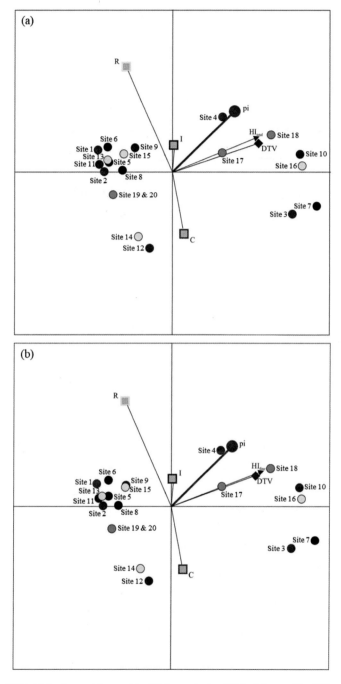

Fig. 4.11 GAIA biplot for: **a** risk posed by HMs attached to TS; **b** risk posed by HMs attached to fine solids; **c** risk posed by PAHs attached to TS; **d** risk posed by PAHs attached to fine solids

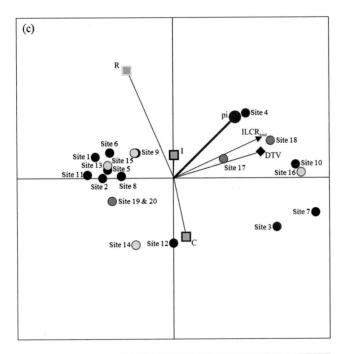

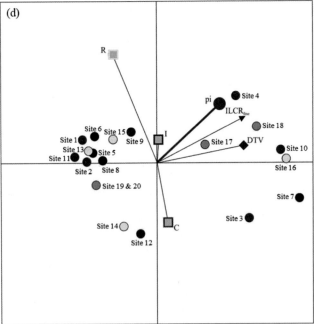

Fig. 4.11 (continued)

Table 4.4 PROMETHEE ranking of combined risk from HMs and PAHs

Rank	Study sites	Suburb
1	Site 10	Suburb 3
2	Site 7	Suburb 2
3	Site 9	Suburb 3
4	Site 12	Suburb 3
5	Site 3	Suburb 1
6	Site 4	Suburb 1
7	Site 11	Suburb 3
8	Site 8	Suburb 2
9	Site 6	Suburb 2
10	Site 16	Suburb 4
11	Site 2	Suburb 1
12	Site 5	Suburb 2
13	Site 1	Suburb 1
14	Site 15	Suburb 4
15	Site 13	Suburb 4
16	Site 14	Suburb 4
17	Site 18	Suburb 5
18	Site 17	Suburb 5
19	Site 19	Suburb 5
19	Site 20	Suburb 5

site. This is expected to result in frequent braking and starting activities, with high build-up load of HMs and PAHs on road surfaces. Consequently, the concentration of HMs and PAHs in stormwater from Site 7 was very high leading to high stormwater risk. Contrary to Site 7, the commercial activities at Site 3 and 4 are mainly office premises, catering, retail, hospitality and tourism (Fig. 4.14). These activities would generate relatively smaller HMs and PAHs loads compared to Site 7. Consequently, the risk posed by these two sites was lower than Site 7. However, these activities can attract relatively high traffic volumes resulting in high stormwater risk at these sites. Additionally, Site 3 and 4 are close to the central business district and there are many traffic lights and parking bays in the vicinity. Hence, traffic congestion is common near the two sites, leading to frequent brake-start activities. High traffic volume and congested traffic are hypothesized to contribute to the high HM and PAH concentrations and risk.

4.4.2.2 Moderate Risk Sites

The nine road sites generating moderate stormwater risk include Site 1, 2, 5, 6, 8, 13, 14, 15 and 16. The anthropogenic activities in the vicinity of Site 1, 2, 5, 6 and 8 are

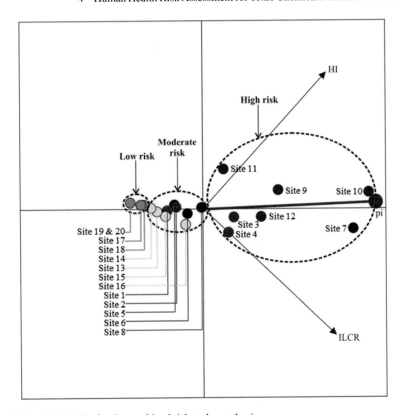

Fig. 4.12 GAIA biplot for the combined risk at the study sites

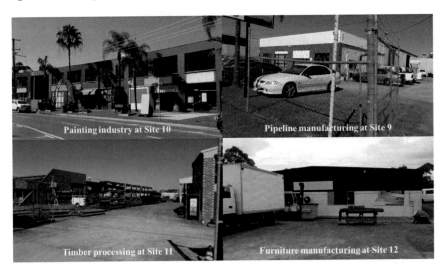

Fig. 4.13 Industrial activities at Site 9, 10, 11 and 12

Fig. 4.14 Commercial activities at Site 7, 3 and 4

a mix of commercial and residential activities and the main commercial activities in the vicinity of these sites consist of education, office premises, hospitality and retail (Fig. 4.15). Compared with the sites with high stormwater risk, the traffic volumes at the moderate risk sites are relatively low. Therefore, traffic generated HMs and PAHs loads at these nine sites are also comparatively low as well as the stormwater risk. Site 8 connects with an arterial road, with the traffic being usually congested and the traffic volume in the surrounding area is high. Therefore, the build-up loads of HMs and PAHs on the road surfaces at Site 8 were high and stormwater risk was also the highest in the nine moderate risk sites (Table 4.4). Site 13, 14, 15 and 16 are in typical residential areas and there are only residential activities at these sites which would generate relatively low pollutant loads and in turn, the associated stormwater risk was low. Compared with Site 1, 2, 5, 6 and 8 where residential areas are mixed with commercial areas, Site 13, 14, 15 and 16 with residential activities alone would exert lower risk due to the less intense anthropogenic activities.

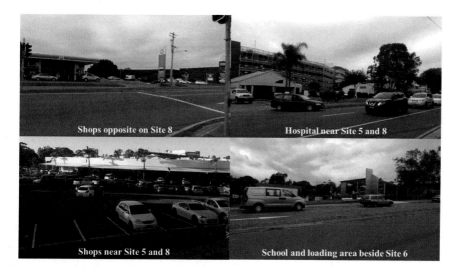

Fig. 4.15 Commercial activities at Site 5, 6 and 8

Fig. 4.16 Land use at Site 17, 18, 19 and 20

4.4.2.3   Low Risk Sites

Site 17, 18, 19 and 20 were identified as having low stormwater risk. These sites are in a natural area where extremely limited anthropogenic activities occur and most of the surrounding land area consists of pervious surfaces (Fig. 4.16). The limited anthropogenic activities are attributed to be the reason for the low pollutant loads and the lowest stormwater risk.

4.4.2.4   Selection of Appropriate Qualitative Parameters

The above discussion confirms that stormwater risk is primarily influenced by qualitative parameters such as type and function of buildings and frequent vehicle brake-start activities. Relevant qualitative parameters for the study sites are summarized in Table 4.5 and their values were obtained through field investigations.

In order to select appropriate qualitative parameters to evaluate human health risk posed by HMs and PAHs in stormwater, the GAIA method was employed to investigate the relationship between risk indices and the qualitative parameters. Two data matrices representing risk posed by HMs and PAHs were prepared for the GAIA analysis. Each matrix consisted of 20 rows (20 study sites as the objects) and seven columns (HI or ILCR, FIS, FVS, FCS, FRH, IP, BSF as variables). Figure 4.17 presents the GAIA analysis result. It is evident from Fig. 4.17a that FVS, BSF, FIS are positively correlated to HI, while FRH is negatively correlated to HI. Besides, angles between IP, FCS and HI are orthogonal which means that these parameters are not correlated and do not influence the risk. Similar results are evident in Fig. 4.17b which depicts the correlation between ILCR and qualitative parameters.

Table 4.5 Traffic and land use related qualitative parameters for the study sites

Study sites	Fraction of industrial premises (FIS)	Fraction of vehicle related businesses (FVS)	Fraction of other commercial businesses (FCS)	Fraction of residential houses (FRH)	Impervious area percentage (IP)	Braking or starting frequency (BSF)
Site 1	0	1	2	2	0.5613	1
Site 2	0	1	2	2	0.5875	1
Site 3	0	1	3	1	0.4922	3
Site 4	0	1	3	1	0.5868	3
Site 5	0	1	2	2	0.7124	1
Site 6	0	1	3	2	0.7601	1
Site 7	0	3	1	1	0.8376	3
Site 8	0	1	3	2	0.7954	3
Site 9	3	3	1	1	0.6567	3
Site 10	3	3	1	1	0.4595	3
Site 11	3	3	1	1	0.6423	3
Site 12	3	3	1	1	0.4425	3
Site 13	0	0	0	3	0.6232	1
Site 14	0	0	0	3	0.3152	1
Site 15	0	0	0	3	0.6329	1
Site 16	0	0	0	3	0.3415	1
Site 17	0	0	0	1	0.0000	1
Site 18	0	0	0	1	0.0000	1
Site 19	0	0	0	1	0.0000	1
Site 20	0	0	0	1	0.0000	1

Note 0, 1, 2 and 3 represent none, small or low, moderate, large or high level of anthropogenic activities

These results indicate that FVS, BSF, FIS and FRH exert strong influence on risk. Considering that FIS and FRH can be represented by the industrial and residential land use area percentage (i.e. quantitative parameters I and R), they were not selected as independent variables to evaluate stormwater risk. Consequently, FVS and BSF were selected for HM and PAH concentrations estimation and risk assessment.

4.5 Further Refinement of the Risk Assessment

Based on the discussion in Sect. 4.4, both, quantitative parameters (DTV, C, I and R) and qualitative parameters (FVS and BSF) were taken into consideration to assess the risk posed by HMs and PAHs in stormwater to human health. Pollutant concentrations

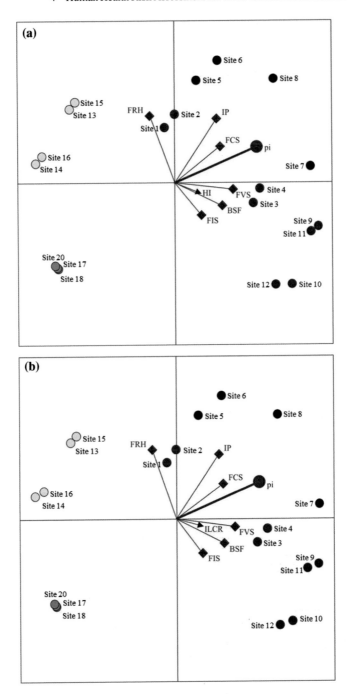

Fig. 4.17 GAIA biplot: **a** based on HI and qualitative parameters; **b** based on ILCR and qualitative parameters

in stormwater from the study sites were initially determined using the R package based on the modelling procedure discussed in Chap. 3. This was followed by refining the risk assessment which was undertaken based on HM and PAH concentrations using the methodology explained in Sect. 4.2. The refined risk assessment equations are given in Eqs. 4.12–4.15 (Ma, 2016) to estimate human health risks posed by HMs and PAHs associated with total and fine solids in stormwater. It can be noted that values of BSF are assigned either 0 and 1 for low and high braking and starting frequency. In terms of FVS, two qualitative parameters, namely FVS_L and FVS_H, are included. Values of FVS_L are assigned either 0 and 1 for none and low braking and starting frequency. Values of FVS_H are assigned either 0 and 1 for none and high braking and starting frequency.

$$
\begin{aligned}
\text{Risk}_{\text{HMtotal}} = {}& 0.05 \times e^{5.2354 - 2.1669 \times \ln\text{DTV} + 0.1885 \times (\ln\text{DTV})^2 + 3.2857 \times C + 3.2755 \times I + 0.9365 \times R - 0.7111 \times \text{BSF}} \\
& + 68.20 \times e^{-8.6167 + 0.2716 \times \ln\text{DTV} + 3.6168 \times C + 4.7310 \times I + 1.1417 \times R - 0.5602 \times \text{FVS\_L}} \\
& + 2.70 \times e^{-5.922 + 0.3220 \times \ln\text{DTV} + 1.598 \times C + 27.5430 \times I - 129.9271 \times I^2 + 0.5266 \times R} \\
& + 0.02 \times e^{-1.7381 + 0.3334 \times \ln\text{DTV} + 4.1151 \times C + 27.5769 \times I - 119.9427 \times I^2 - 0.2766 \times \text{FVS\_L} - 0.5848 \times \text{BSF}} \\
& + 1.33 \times e^{-8.4381 + 0.1820 \times \ln\text{DTV} + 1.9328 \times C + 7.7062 \times I + 0.8789 \times R + 1.6058 \times \text{FVS\_H}} \\
& + 0.63 \times e^{-8.843 + 0.5132 \times \ln\text{DTV} + 5.5407 \times C + 3.0988 \times I + 0.6813 \times R + 1.2059 \times \text{FVS\_H} - 0.5906 \times \text{BSF}} \\
& + 0.10 \times e^{-7.2871 + 0.4862 \times \ln\text{DTV} + 4.1299 \times C + 52.7886 \times I - 236.2639 \times I^2 + 1.0970 \times R - 0.4976 \times \text{BSF}} \\
& + 409.18 \times e^{-15.0754 + 0.7340 \times C + 6.6554 \times I - 1.4475 \times R + 0.2505 \times \text{FVS\_L} - 0.4857 \times \text{BSF}} \\
& + 10.83 \times e^{-8.4649 + 0.4781 \times \ln\text{DTV} + 7.4643 \times C + 51.4039 \times I - 210.5399 \times I^2 - 1.2029 \times \text{FVS\_L} - 1.2544 \times \text{FVS\_H} - 0.4706 \times \text{BSF}}
\end{aligned}
\tag{4.12}
$$

$$
\begin{aligned}
\text{Risk}_{\text{HMfiner}} = {}& 0.05 \times e^{\substack{-6.8022 + 0.6341 \times \ln\text{DTV} + 2.9511 \times C + 30.5369 \times I - 140.5479 \times I^2 + 1.6493 \\ \times R - 0.7633 \times \text{FVS}_L - 0.6815 \times \text{BSF}}} \\
& + 68.20 \times e^{-25.0601 + 4.3165 \times \ln\text{DTV} - 0.2727 \times (\ln\text{DTV})^2 + 0.7033 \times R + 1.5514 \times FVS_H} \\
& + 2.70 \times e^{\substack{-21.0598 + 3.8874 \times \ln\text{DTV} - 0.2336 \times (\ln\text{DTV})^2 + 25.2073 \times I - 128.1087 \times I^2 + 1.2651 \times R \\ -0.5441 \times FVS_L + 0.6172 \times FVS_H}} \\
& + 0.02 \times e^{\substack{-13.9268 + 2.9694 \times \ln\text{DTV} - 0.1704 \times (\ln\text{DTV})^2 + 3.2864 \times C + 44.3354 \times I - 205.3793 \times I^2 + 1.2523 \times R \\ -0.8872 \times FVS_L - 0.4395 \times BSF}} \\
& + 1.33 \times e^{-24.6988 + 4.2912 \times \ln\text{DTV} - 0.2863 \times (\ln\text{DTV})^2 + 7.1189 \times I + 3.9997 \times R - 5.3312 \times R^2 + 2.1279 \times FVS_H} \\
& + 0.63 \times e^{-23.3460 + 4.1531 \times \ln\text{DTV} - 0.2487 \times (\ln\text{DTV})^2 + 2.4809 \times C + 0.9682 \times R + 1.6005 \times FVS_H} \\
& + 0.10 \times e^{-17.2335 + 2.7326 \times \ln\text{DTV} - 0.1466 \times (\ln\text{DTV})^2 + 2.0193 \times R + 1.4526 \times FVS_H} \\
& + 409.18 \times e^{-24.1876 + 2.8568 \times \ln\text{DTV} - 0.1491 \times (\ln\text{DTV})^2 - 1.2810 \times C} \\
& + 10.83 \times e^{\substack{-12.2442 + 0.7574 \times \ln\text{DTV} + 26.3351 \times C - 32.8041 \times C^2 + 57.7396 \times I - 243.3053 \times I^2 \\ +1.2913 \times R - 3.9695 \times FVS_L - 2.9992 \times FVS_H - 0.6539 \times BSF}}
\end{aligned}
\tag{4.13}
$$

$$
\begin{aligned}
\text{Risk}_{\text{PAHtotal}} = {}& \frac{33,2059.3}{1,788,500} \\
& \times 0.001 \times e^{\substack{-13.9375 + 0.1576 \times \ln\text{DTV} + 6.0175 \times C + 68.6756 \times I - 311.5474 \times I^2 + 2.4776 \times R - 0.9163 \\ \times \text{FVS}_L - 1.1478 \times \text{FVS}_H - 0.5941 \times \text{BSF}}} \\
& + 0.001 \times e^{\substack{-19.8234 + 0.8016 \times \ln\text{DTV} + 6.0153 \times C + 54.5872 \times I - 235.5529 \times I^2 + 2.4933 \times R - 1.0615 \\ \times \text{FVS}_L - 1.3930 \times \text{FVS}_H - 0.9856 \times \text{BSF}}} \\
& + 0.001 \times e^{\substack{-17.7646 + 0.4760 \times \ln\text{DTV} + 4.4852 \times C + 37.5646 \times I - 153.9882 \times I^2 + 1.0939 \times R - 0.8900 \\ \times \text{FVS}_L - 0.9199 \times \text{FVS}_H - 1.1782 \times \text{BSF}}} \\
& + 0.001 \times e^{\substack{-15.5124 + 0.3572 \times \ln\text{DTV} + 4.3171 \times C + 47.3834 \times I - 201.9721 \times I^2 + 1.3830 \times R - 1.4614 \\ \times \text{FVS}_L - 1.4913 \times \text{FVS}_H - 0.3870 \times \text{BSF}}}
\end{aligned}
$$

$$+ 0.001 \times e^{-13.5478+0.3597\times\ln DTV+2.2984\times C+29.7695\times I-124.8451\times I^2+0.9774\times R-0.8053\times BSF}$$

$$+ 0.01 \times e^{-16.4012+0.4909\times\ln DTV+2.9089\times C+30.1952\times I-120.4977\times I^2-1.0816\times BSF}$$

$$+ 0.001 \times e^{\substack{-9.4698-1.1452\times\ln DTV+0.1166\times(\ln DTV)^2+4.1308\times C+29.8425\times I-128.2472\times I^2+1.1498 \\ \times R-1.3030\times FVS_L-1.0456\times BSF}}$$

$$+ 0.001 \times e^{-14.6646+0.5522\times\ln DTV+2.4984\times C+2.8381\times I+1.6270\times R-0.4150\times FVS_L-0.6512\times BSF}$$

$$+ 0.1 \times e^{\substack{-19.2122+0.8082\times\ln DTV+5.2110\times C+46.7275\times I-208.7814\times I^2+4.9091\times R-4.9529\times R^2-0.8750 \\ \times FVS_L-1.1090\times FVS_H-0.7659\times BSF}}$$

$$+ 0.01 \times e^{-17.0734+0.7185\times\ln DTV+3.5279\times C+29.7688\times I-131.1462\times I^2+0.8078\times R-0.7018\times BSF}$$

$$+ 0.1 \times e^{-17.8669+0.7511\times\ln DTV+3.4531\times C+25.2337\times I-115.0348\times I^2+1.1233\times R-0.6657\times BSF}$$

$$+ 1 \times e^{-18.8690+0.8781\times\ln DTV+2.2772\times C+1.4908\times R}$$

$$+ 0.1 \times e^{\substack{-17.8037+0.7329\times\ln DTV+6.6727\times C+52.7853\times I-238.0639\times I^2+1.6387\times R-1.1872 \\ \times FVS_L-1.2115\times FVS_H-0.7137\times BSF}}$$

$$+ 5 \times e^{\substack{-18.2572+0.7225\times\ln DTV+8.3825\times C+72.1128\times I-314.2562\times I^2+1.5766\times R-1.0057 \\ \times FVS_L-1.4935\times FVS_H-1.3597\times BSF}}$$

$$+ 0.01 \times e^{-15.7296+0.5912\times\ln DTV+1.4408\times C+45.0252\times I-202.8541\times I^2+2.2230\times R-0.5179\times BSF} \tag{4.14}$$

$$\text{Risk}_{\text{PAHfiner}} = \frac{332{,}059.3}{1{,}788{,}500}$$

$$\times 0.001 \times e^{-16.4869+0.3607\times\ln DTV+1.7239\times C+1.5317\times I+2.8164\times R-0.7319\times FVS_L}$$

$$+ 0.001 \times e^{\substack{-33.2872+4.2847\times\ln DTV-0.2325\times(\ln DTV)^2-5.9950\times C+14.7871\times C^2+4.6203 \\ \times I+1.9302\times R-0.4105\times BSF}}$$

$$+ 0.001 \times e^{\substack{-29.0043+3.1518\times\ln DTV-0.1662\times(\ln DTV)^2+8.8727\times C-11.6449\times C^2+5.3196\times I+0.8567 \\ \times R-1.8578\times FVS_L-1.1695\times FVS_H-1.2009\times BSF}}$$

$$+ 0.001 \times e^{\substack{-18.4995+0.6292\times\ln DTV+3.4958\times C+7.0867\times I+1.8015\times R-2.5659 \\ \times FVS_L-2.2351\times FVS_H+0.4647\times BSF}}$$

$$+ 0.001 \times e^{-16.2627+0.5957\times\ln DTV+2.7675\times I+1.7208\times R-0.4207\times FVS_L-0.3398\times BSF}$$

$$+ 0.01 \times e^{-19.3564+0.8000\times\ln DTV+1.9097\times C+5.5251\times I-0.3824\times FVS_L-1.1861\times BSF}$$

$$+ 0.001 \times e^{\substack{-18.1027+0.8260\times\ln DTV+2.8339\times C+4.2088\times I-2.9652\times R+6.8062 \\ \times R^2-1.5358\times FVS_L-0.6618\times BSF}}$$

$$+ 0.001 \times e^{-17.1744+0.7539\times\ln DTV+1.2404\times C+2.1712\times R-0.7168\times FVS_L-0.3770\times BSF}$$

$$+ 0.1 \times e^{\substack{-32.1769+4.0290\times\ln DTV-0.2054\times(\ln DTV)^2+1.8259\times C+2.3244\times I+1.4597 \\ \times R-0.5785\times FVS_L-0.5054\times BSF}}$$

$$+ 0.01 \times e^{-20.4228+1.0262\times\ln DTV+1.6406\times C+3.3731\times I+1.5066\times R-0.4627\times BSF}$$

$$+ 0.1 \times e^{-30.2113+3.8281\times\ln DTV-0.2036\times(\ln DTV)^2+1.2840\times C+1.6044\times R-0.5705\times FVS_L}$$

$$+ 1 \times e^{-29.7000+3.4408\times\ln DTV-0.1608\times(\ln DTV)^2+0.8037\times C+1.4459\times R}$$

$$+ 0.1 \times e^{-26.6484+2.7348\times\ln DTV-0.1251\times(\ln DTV)^2+1.5743\times C+1.9833\times R-0.9597\times FVS_L}$$

$$+ 5 \times e^{\substack{-30.2300+3.5487\times\ln DTV-0.1745\times(\ln DTV)^2+18.6129\times C-18.5837\times C^2+66.3846\times I-283.1660\times I^2 \\ +1.8129\times R-2.7107\times FVS_L-2.5458\times FVS_H-1.5929\times BSF}}$$

$$+ 0.01 \times e^{-25.9924+3.0257\times\ln DTV-0.1545\times(\ln DTV)^2-2.5337\times C+29.9913\times I-131.4012\times I^2+2.4693\times R} \tag{4.15}$$

4.6 Conclusions

This chapter presents a risk assessment modelling approach to estimate human health risks posed by HMs and PAHs associated with total and fine solids present in stormwater. A series of mathematical equations were developed for risk assessment by incorporating quantitative parameters (daily traffic volume and land use area percentages) and qualitative parameters (fraction of motor vehicle related businesses and braking-starting frequency) relevant to traffic and land use. These equations will enable the estimation of stormwater risk based on traffic and land use characteristics. Furthermore, they would contribute to more reliable stormwater risk management based on appropriate traffic and land use planning in urban areas. Although these predictive equations are only applicable within their specified limits, they provide a robust approach to stormwater risk assessment and can be modified to suit different geographic locations.

References

ANZECC. (2000). Australian and New Zealand guidelines for fresh and marine water quality. Canberra, Australian and New Zealand Environment and Conservation Council and Agriculture and Resource Management Council of Australia and New Zealand, pp. 1–103.

Fabian, P. S., Lee, D. H., Shin, S. W., & Kang, J.-H. (2022). Assessment of pyrene adsorption on biochars prepared from green infrastructure plants: Toward a closed-loop recycling in managing toxic stormwater pollutants. *Journal of Water Process Engineering, 48*, 102929.

Gbeddy, G., Egodawatta, P., Goonetilleke, A., Ayoko, G., & Chen, L. (2020). Dataset for the quantitative structure-activity relationship (QSAR) modeling of the toxicity equivalency factors (TEFs) of PAHs and transformed PAH products. *Data in Brief, 28*, 104821.

IRIS (2015, Tuesday, 17 March 2015). A-Z List of Substances. Retrieved 02 April, 2015, 2015, from https://www.epa.gov/iris

Kho, Y., Lee, E.-H., Chae, H. J., Choi, K., Paek, D., & Park, S. (2014). 1-Hydroxypyrene and oxidative stress marker levels among painting workers and office workers at shipyard. *International Archives of Occupational and Environmental Health, 88*, 297–303.

Ma, Y. (2016). *Human health risk of toxic chemical pollutants generated from traffic and land use activities.* Queensland University of Technology.

Ma, Y., Deilami, K., Egodawatta, P., Liu, A., McGree, J., & Goonetilleke, A. (2019). Creating a hierarchy of hazard control for urban stormwater management. *Environmental Pollution, 255*, 113217.

Ma, Y., Egodawatta, P., McGree, J., Liu, A., & Goonetilleke, A. (2016). Human health risk assessment of heavy metals in urban stormwater. *Science of the Total Environment, 557–558*, 764–772.

Ma, Y., Liu, A., Egodawatta, P., McGree, J., & Goonetilleke, A. (2017a). Assessment and management of human health risk from toxic metals and polycyclic aromatic hydrocarbons in urban stormwater arising from anthropogenic activities and traffic congestion. *Science of the Total Environment, 579*, 202–211.

Ma, Y., Liu, A., Egodawatta, P., McGree, J., & Goonetilleke, A. (2017b). Quantitative assessment of human health risk posed by polycyclic aromatic hydrocarbons in urban road dust. *Science of the Total Environment, 575*, 895–904.

Ma, Y., McGree, J., Liu, A., Deilami, K., Egodawatta, P., & Goonetilleke, A. (2017c). Catchment scale assessment of risk posed by traffic generated heavy metals and polycyclic aromatic hydrocarbons. *Ecotoxicology and Environmental Safety, 144,* 593–600.

NHMRC&AWRC. (2011). *Australian drinking water guidelines.* Australian Government Publishing Service.

Nisbet, I. C., & LaGoy, P. K. (1992). Toxic equivalency factors (TEFs) for polycyclic aromatic hydrocarbons (PAHs). *Regulatory Toxicology and Pharmacology, 16*(3), 290–300.

Patel, A. B., Mahala, K., Jain, K., & Madamwar, D. (2018). Development of mixed bacterial cultures DAK11 capable for degrading mixture of polycyclic aromatic hydrocarbons (PAHs). *Bioresource Technology, 253,* 288–296.

USEPA (1989). *Risk assessment guidance for superfund, Volume I, human health evaluation manual (Part A).* US Environmental Protection Agency.

Chapter 5
Implications for Stormwater Management and Recommendations for Future Research

Abstract This chapter summarizes the major outcomes derived from the research study which aimed to assess the human health risk associated with urban stormwater as influenced by traffic and land use parameters and the practical application of the research outcomes to urban stormwater management. Recommendations are provided to strengthen the hierarchy of hazard control strategies to minimize stormwater risk and current water reuse guidelines. This chapter also discusses the current knowledge gaps and provides recommendations for future research.

Keywords Stormwater management · Risk management strategies · Water reuse guidelines · Hierarchy of hazard control · Urban traffic · Land use · Stormwater quality

5.1 Background

Urban stormwater pollution can exert toxic impacts on human health through stormwater reuse (Wijesiri et al., 2020). Therefore, appropriate management of stormwater risk is essential in order to utilize stormwater resources (Ma et al., 2019). In this context, designing effective control measures to minimize urban stormwater risk is required. The study developed risk assessment models by establishing the mathematical relationships between stormwater risk and influential factors such as traffic and land use characteristics. These risk assessment models can be applied for formulating management strategies to improve urban stormwater quality and to reduce the risk associated with stormwater reuse.

This chapter summarizes the major outcomes derived from the previous chapters and discusses their practical applications for urban stormwater management and reuse. Furthermore, current knowledge gaps in the context of this project are identified and recommendations are provided for future research.

© The Author(s), under exclusive license to Springer Nature Singapore Pte Ltd. 2023
Y. Ma et al., *Human Health Risk Assessment of Toxic Chemical Pollutants in Stormwater*,
SpringerBriefs in Water Science and Technology,
https://doi.org/10.1007/978-981-19-9616-0_5

5.2 Implications for Urban Stormwater Management

5.2.1 Implications Related to Current Water Quality Guidelines

Major analytical outcomes derived from this project and their potential practical implications in relation to current water quality guidelines are discussed below.

- Understanding the overall human health risk associated with HMs and PAHs for appropriate stormwater reuse is a critical requirement. Evaluation of individual HM and PAH concentrations in stormwater cannot provide this required level of understanding as robust methodology is not available to translate pollutant concentrations to the risk posed to human health.
- Taking into consideration the combined toxic effects of multiple HMs and PAHs is essential for effective water quality management. Even though an individual HM and PAH may not be toxic to human health, multiple HM and PAH mixtures can pose risk to human health.
- Further, water quality guidelines need to provide safe threshold levels for HMs and PAHs for different stormwater reuse purposes. Stormwater is a potential water resource for potable and recreational use to relieve the ever escalating water shortages being experienced worldwide. People can come into contact with stormwater through various pathways such as consumption and recreational use.

5.2.2 Factors Influencing Stormwater Risk

Stormwater risk is dependent on traffic (daily traffic volume and braking and starting frequency) and land use (commercial, residential and industrial land use area percentages and fraction of motor vehicle related businesses) characteristics. Therefore, stormwater risk assessment and risk management should consider these factors. Key outcomes derived from this project was adopted for creating a novel stormwater risk assessment approach. The practical implications of the research findings are as follows:

- The land use types need to be initially identified for stormwater risk assessment. Industrial activities contribute the highest human health risk, followed by commercial and residential activities.
- Identification of land use characteristics is also needed. Commercial activities can be categorized into two types as: motor vehicle related activities and other commercial activities such as offices, education, retail, catering and hospitality. Residential areas also contain two types as: residential areas mixed with commercial areas and solely residential areas. Motor vehicle related activities can generate

relatively high stormwater risk compared to other commercial activities. Residential areas mixed with commercial areas can result in relatively high risk compared to only residential areas.

- Daily traffic volume needs to be determined. For a specific land use type, stormwater risk generally increases with increasing traffic volume.
- Traffic characteristics in the vicinity need to be identified. Frequent braking and starting due to traffic congestion and the presence of parking areas can result in high risk associated with stormwater.

Accordingly, a generic roadmap considering the influential factors was developed for assessing the stormwater risk to human health posed by HMs and PAHs (Fig. 5.1). The roadmap consists of four key influential factors taken into consideration, namely, traffic volume, land use, traffic characteristics and land use characteristics. Based on the roadmap, a robust approach for the assessment and management of stormwater risk was proposed as presented as a flowchart in Fig. 5.2. It is important to note that the roadmap and flowchart given in Figs. 5.1 and 5.2, respectively, are generic and can be applied in any urban area in other geographic locations. Practical strategies for stormwater risk management as presented in Fig. 5.2 are discussed below.

The hierarchy of hazard control procedure has been widely adopted for risk management in various domains such as disaster management, occupational hygiene and food safety (Ma et al., 2020; McLeod & Curtis, 2022). Therefore, this approach can be equally effective in the management of human health risks posed by HMs and PAHs in urban stormwater. Hierarchy of hazard control is a sequence of actions to eliminate or reduce risk. It consists of five components: elimination, substitution, engineering control, administrative control and use of personal protective equipment. The effectiveness of control of the above sequence is in descending order, while the

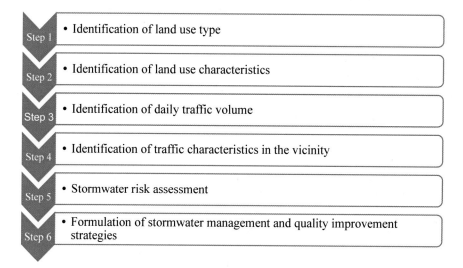

Fig. 5.1 Roadmap to assess and manage stormwater risk

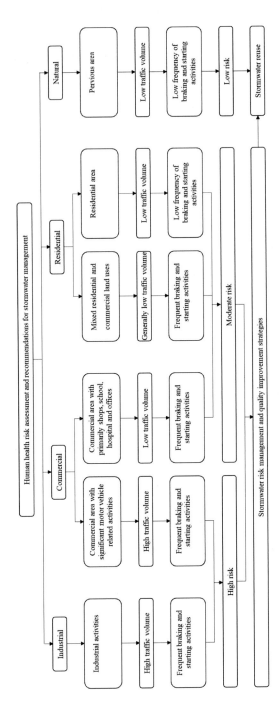

Fig. 5.2 Flowchart for human health risk assessment and management of HMs and PAHs in stormwater

cost to business is also in the descending order. Therefore, judgement is required for selecting appropriate strategies, taking efficiency and cost into consideration. Major analytical outcomes obtained from this project can be adopted to justify hazard control strategies and the potential practical implications in relation to stormwater risk management are discussed below.

- Elimination: Elimination of Cr and heavy PAHs is the most effective approach to minimize human health risk associated with urban stormwater. Cr and heavy PAHs were identified as the most toxic HM and PAH species in urban stormwater contributing the largest fraction of risk. However, it is not practical to completely eliminate all Cr and heavy PAHs present in stormwater. Hence, reducing the generation of typical pollutants in urban areas is critical. For example, reducing traffic volume can effectively control the generation of Cr and heavy PAHs on road surfaces because traffic activities are a major source of HMs and PAHs. Besides, road cleaning at motor vehicle related businesses such as car repair facilities and fuel stations need to be strengthened to minimize the transportation of HMs and PAHs from build-up to stormwater runoff.
- Substitution: Instead of eliminating Cr and heavy PAHs, substitution in products that generate Cr and heavy PAHs by alternative materials and technologies should be considered. For example, frequent braking can generate large amounts of HMs and PAHs. Therefore, alternative brake pad materials and fuel without Cr and heavy PAHs being present is needed. Besides, activities such as metal industries that generate large amounts of Cr should be substituted by cleaner production techniques such as nanotechnology approaches.
- Engineering control: Appropriate engineering control methods can be applied in the priority areas with high human health risk identified in the risk map. For example, water sensitive urban design (WSUD) is widely adopted for urban stormwater management. Accordingly, stormwater quality in the priority areas with high stormwater risk can be improved through the use of WSUD technologies such as improved filter media focusing on removing Cr and heavy PAHs. This in turn implies that the adoption of localized WSUD systems rather than the use of centralized systems would improve the efficacy of stormwater treatment. In addition to WSUD, stormwater reuse can be adopted in the priority areas for non-potable use, such as in ornamental ponds, toilet flushing, road cleaning and building cooling, to avoid people coming into contact with stormwater.
- Administrative control: Administrative control is mainly aimed at warning people about the risk associated with stormwater. For example, signage can be provided beside a stormwater ornamental pond in a priority area to warn people that the water is harmful to health. Furthermore, since stormwater is a primary source of pollutants in urban receiving waters, signage could be installed beside outfalls into the receiving waters in the priority areas to warn people not to swim or make contact immediately after rainfall events.
- Personal Protective Equipment: This control strategy is suggested to be undertaken by people who are most likely to come into contact with stormwater in

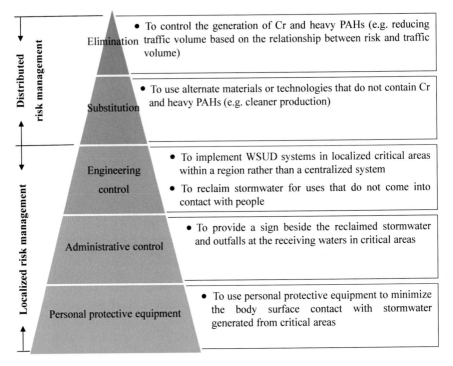

Fig. 5.3 Hierarchy of hazard control for stormwater risk (Ma et al., 2019)

priority areas. For example, diving suits and goggles should be used by workers to minimize the skin surface area likely to come into contact with stormwater.

Based on the discussion above, hierarchy of hazard control for stormwater risk is illustrated in Fig. 5.3. Elimination and substitution can be applied for distributed risk management at macro scale, while engineering control, administrative control and personal protective equipment are suitable for localized risk management at micro scale.

5.3   Recommendations for Future Research

This research project contributes new knowledge by assessing human health risk arising from HMs and PAHs in stormwater. Moreover, the influence of traffic and land use on stormwater risk was investigated in-depth and the linkage between risk and traffic and land use relevant parameters was mathematically established. However, several significant knowledge gaps still remain and need further investigation.

5.3.1 Risk Assessment of Toxic Pollutants to Receiving Waters

This project focused on risk assessment of toxic chemical pollutants in stormwater. Stormwater is a major source of pollutants to receiving waters and people can come into contact with pollutants in stormwater through exposure to urban receiving waters. Therefore, assessing the risk from urban receiving waters is also critical. Due to the dilution of pollutant concentrations in stormwater discharged into receiving waters, the dilution factors need consideration for risk assessment. The outcomes derived from this project provide the foundational step for assessing the risk to urban receiving waters.

5.3.2 Risk Assessment of Other Toxic Chemical Pollutants in Urban Stormwater

This research study evaluated the toxic effects of HMs and PAHs in urban stormwater. However, the presence of many other pollutants due to anthropogenic activities common to urban areas have been identified which are referred to as emerging contaminants. The presence of emerging contaminants is increasingly recognized in urban stormwater and some can be extremely toxic to human health. Therefore, further research is recommended to assess the risk arising from emerging contaminants for safe stormwater reuse. The robust risk assessment approach developed in this study provides the foundation for evaluating risk from emerging contaminants.

5.3.3 Assessing Ecological Risk Associated with Urban Stormwater

This study only discussed human health risk associated with urban stormwater. However, toxic pollutants present in stormwater can also lead to ecological risk through stormwater reuse such as irrigation and aquifer recharge. Therefore, the investigation of the impact of toxic chemical pollutants on ecological health is recommended.

References

Ma, B., Han, Y., Cui, S., Geng, Z., Li, H., & Chu, C. (2020). Risk early warning and control of food safety based on an improved analytic hierarchy process integrating quality control analysis method. *Food Control, 108*, 106824.

Ma, Y., Deilami, K., Egodawatta, P., Liu, A., McGree, J., & Goonetilleke, A. (2019). Creating a hierarchy of hazard control for urban stormwater management. *Environmental Pollution, 255,* 113217.

McLeod, S., & Curtis, C. (2022). Integrating urban road safety and sustainable transportation policy through the hierarchy of hazard controls. *International Journal of Sustainable Transportation, 16*(2), 166–180.

Wijesiri, B., Liu, A., & Goonetilleke, A. (2020). Impact of global warming on urban stormwater quality: From the perspective of an alternative water resource. *Journal of Cleaner Production, 262,* 121330.

Index

© The Author(s), under exclusive license to Springer Nature Singapore Pte Ltd. 2023
Y. Ma et al., *Human Health Risk Assessment of Toxic Chemical Pollutants in Stormwater*,
SpringerBriefs in Water Science and Technology,
https://doi.org/10.1007/978-981-19-9616-0

Printed in the United States
by Baker & Taylor Publisher Services